THE LEAD MINERS
OF THE NORTHERN PENNINES
in the eighteenth and nineteenth centuries

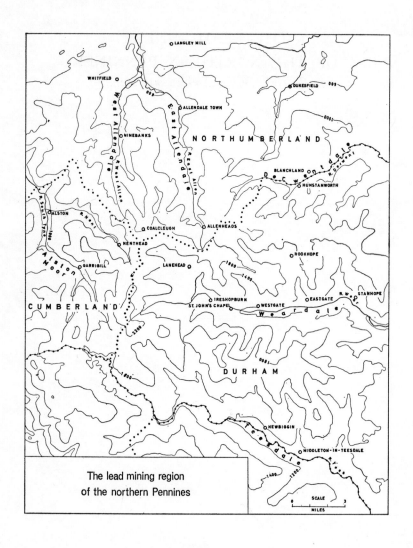

The lead mining region
of the northern Pennines

THE LEAD MINERS
OF THE
NORTHERN PENNINES

in the eighteenth and nineteenth centuries

C. J. Hunt

Manchester University Press

© 1970 Christopher John Hunt

Published by the University of Manchester at
The University Press
316–324 Oxford Road, Manchester, M13 9NR

UK standard book number: 7190 0380 6

Printed in Great Britain by
Butler & Tanner Ltd, Frome and London

Contents

Illustrations

Acknowledgements

To the following I record my thanks for help in the writing of this book, which started life as a thesis for the Durham University M.Litt. degree: all the private owners, archivists and librarians who allowed me access to books and papers; Mr John Knipe and his staff at the Atlas of Northern England drawing office, Newcastle University, for help with the general map of the mining region; Dr W. H. Chaloner, Dr Kingsley Dunham, Dr and Mrs Angus Lunn and Dr T. C. Smout, all of whom read and criticised parts of the manuscript; Dr Mark Hughes for allowing me to quote from his own work on lead mining and for many helpful suggestions about sources; my wife and my mother, both of whom typed and retyped successive drafts; and especially Emeritus Professor Edward Allen who helped me at every stage of its writing.

ABBREVIATIONS

B/B Blackett/Beaumont MSS.
L.L.C. London Lead Company MSS.
Adm. Admiralty: Greenwich Hospital MSS.
G.H. (Wigan) MS. Greenwich Hospital (Wigan) MS.
1842 report. Children's Employment (Mines), R. Com., 1842.
1864 report. Mines Commission, 1864.

I

Introduction

This work is a study of the social and economic conditions of lead miners in the northern Pennines during the eighteenth and nineteenth centuries. In this period occurred an industrial and managerial revolution that greatly changed the lives of the miners, outside work as well as during working hours. The loose industrial organisation of the eighteenth century, when miners sold the produce of their labour rather than their labour itself, gave place to tight managerial control, not only over the place of work, but also over the miners' entire environment.

The region under consideration is an area of some 650 square miles in the highest and most northerly part of the Pennine hills. Its centre is not far from where the three counties of Northumberland, Cumberland and Durham meet. The region is virtually in the centre of Britain—in fact, Allendale Town claims to be *the* centre. It is drained by six rivers and their tributaries—the South Tyne, the West and East Allen, the Derwent, the Wear and the Tees. The dales of these rivers, in which lie the settlements, are themselves nearly all more than 700 ft above sea level. The ridges between the dales range from 1,500 to 3,000 ft.

The climate is severe by English standards. Winter starts early and continues late, commonly extending into April. The summers tend to be cool and wet. Rain falls on an average of more than 200 days a year. Snow frequently covers the higher hills for more than five months of the year.

Human settlements are limited to the dales, becoming fewer and more scattered as altitude increases and shelter diminishes. In the period 1750–1850 there was an expansion of population and of the number of settlements, which crept nearer and nearer to the heads of the dales. New villages developed, often around the headquarters of mining companies, but the older villages continued to serve as the marketing centres of their respective dales. Alston was the centre for

South Tynedale and the Nent valley (Alston Moor); Allendale Town, for East and West Allendale; Blanchland, for Derwentdale; Stanhope, for Weardale; and Middleton, for Teesdale.

The high moorland which separated dale from dale formed a barrier isolating the inhabitants from their neighbours. The social links of each dale were with those who lived in the areas further down the rivers, rather than with those of the other dales. The region as a whole was orientated towards the east. Alston is in Cumberland, but separated from the rest of the county by the highest hills in the Pennines. Its natural links, therefore, were with Northumberland, along the course of the South Tyne.

In the eighteenth century, roads within the region were appallingly bad, even by contemporary standards. Rain or snow made them virtually impassable. In 1735 a surveyor of the Greenwich Hospital, which had just been given the confiscated northern estates of the Jacobite earls of Derwentwater, requested that he should be allowed to postpone his survey of Alston Moor because 'in the first place it is in such a part of the world that they are seldome without rains, in the next it is so Mountainious and Rotten that it would be with difficulty that a man could walk upon the mosses in many places.' The roads were so bad that carts were not used in the eighteenth century; galloways—trains of packhorses—were the only form of transport. Roads were so poorly constructed that after a few years' use they were often abandoned for new routes, their surfaces having deteriorated to a state worse than that of the surrounding moorland. John McAdam reported to the Greenwich Hospital in 1823 that the roads 'are altogether the worst that have yet come to my knowledge—not only have the old defective methods been followed in the formation, stoning and subsequent repair of the Roads, but the work has been executed in the most slovenly careless manner, without method . . . No pains having been taken to preserve the Roads from the Winter floods, by keeping open the waterways, they are washed out so as to present the appearance of a bed of rocks rather than an artificial road.'[2]

At the beginning of the nineteenth century the proprietors of land and mines realised that the construction of better roads would be a worthwhile investment. In the ten years between 1820 and 1830 were built most of those which exist today (see Fig. 1). The Greenwich Hospital and the two largest mining concerns, the London Lead Company and the Beaumonts, spent thousands of pounds on roads in this period, concentrating on the ones leading out of the dales. The

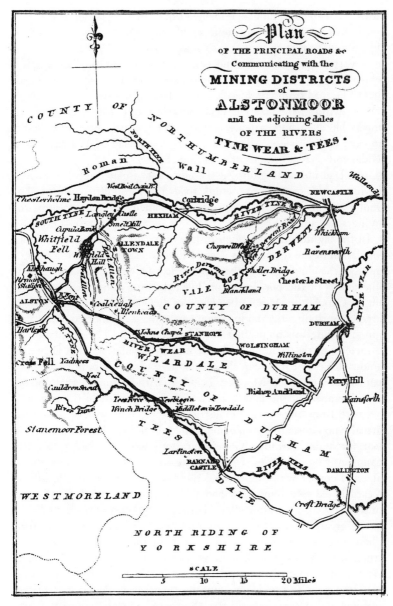

Fig. 1   The lead mining region, showing lines of communication with the east coast in 1833. *From T. Sopwith*, Alston Moor, *1833*

roads between dale and dale remained bad until after the middle of the century. Thomas Sopwith, mine agent and engineer, wrote in 1838 that 'Stanhope is at all times nearly inaccessible from Middleton except by a very circuitous route and particularly so after the heavy rains which have fallen of late.' In the winter of 1846 he similarly noted that 'The road from Allenheads to Coalcleugh [the two major mining centres within Allendale] is deplorably bad and in several places is impassable even with a pony.'³ After 1850 railway lines were built up each dale. Alston, Allendale Town, Wearhead in Weardale and Middleton were all the termini of branch lines.

The lead mining region was therefore very isolated until well into the nineteenth century. No trunk roads ran through it; no traveller went there accidentally, or *en route* for somewhere else. Visitors or immigrants came deliberately.

In the eighteenth and nineteenth centuries the greater part of the population of the upper dales depended upon lead mining for a living. In 1857 the chief agent of the London Lead Company reckoned that nine-tenths of the population of Teesdale were connected with the mines. On Alston Moor 'the whole of the residents ... are, either directly or indirectly, taking them as a mass, connected with the mines.' There was no distinct farming population. 'The population is so mixed up, the farming with the mining population, that they are almost all as one; it is scarcely possible to go into a family occupying half an acre in Alston Moor or Teesdale, without finding that one or more members of the family are workmen employed in the mines.'⁴

Apart from the service industries—shops, inns, etc.—there was virtually no alternative employment to lead mining. The land was too poor for arable farming. Communications were too bad to attract manufacturing industries. In the nineteenth century small deposits of zinc and copper were mined, and the discovery of iron ore in lower Weardale led to a temporary boom in employment there that lasted for some two decades after 1850. However, for most of the period and for the greater part of the region, lead mining was the only industry. There was no employment for women in the nineteenth century, although in the eighteenth some were employed in dressing lead ore. Many lead mining families possessed smallholdings, but these were too small and poor to provide a living by themselves. When lead prices collapsed and mines closed in the 1870's and 1880's the greater part of the population was forced to migrate.

Lead miners were not engaged only in extracting metal from the ground. The produce from the ground was lead *ore*; it had to be dressed ('washed' in local terminology) and smelted before it became lead. The labour force was divided into three parts: the miners proper, the washers, and the smelters. In the eighteenth century the first two divisions were confused, the miners commonly washing their own ore; the smelters, however, were separate. In the nineteenth century a threefold division of the labour force became more rigid in all the larger mining concerns.[5] In addition there were the 'carriers' engaged in the transport of the ore and lead; most of these latter lived outside the mining region.

The mines were worked either directly by the landowner who owned the mining rights, or leased by him to concessionaries. In fact only the Blackett/Beaumont family, who owned the mining rights in Allendale, chose the first alternative. Other major landowners preferred to lease their rights in exchange for cash (as did the Bishop of Durham in Weardale) or a percentage of the ore raised (as was done by the Greenwich Hospital on Alston Moor).[6]

The mining region comprised six distinct areas: Alston Moor; East Allendale, or Allenheads; West Allendale, or Coalcleugh; Derwent; Weardale; Teesdale. In addition there was a small mining field in the Tyne valley, north of Haydon Bridge and Hexham, separated by some seven miles of country devoid of lead deposits from the mining region proper.

Alston Moor was given to Greenwich Hospital in 1735, and the mines, which had not been worked for many years, were gradually opened up. The Hospital did not work them directly but leased them in large and small lots to others in exchange for a percentage of the ore raised. This ore was then sold back to the mining companies until 1767, when the Hospital built a mill at Langley to smelt the ore itself.

The most important lessee of the Greenwich Hospital was the London Lead Company (known also as the 'Governor and Company' or 'the Quaker Company'). It was the largest single employer of labour on Alston Moor from the mid-eighteenth century until the end of the nineteenth. The company operated from two centres, at Nenthead on the river Nent, a tributary of the South Tyne, and on the South Tyne itself at Garrigill.[7] The other mining concerns on Alston Moor were very much smaller than the Lead Company and often had only a transitory existence. In 1824, for example, of the thirty-eight companies leasing mines from the Greenwich Hospital twenty-nine of them employed fewer than twenty hands, and twenty-one fewer than ten. The Hudgill Burn and Rodderup Fell companies were of importance during the nineteenth century. Little evidence survives about the others.

The mines in Allendale were worked directly by the Lord of the Manor, the most important landowner. This title was held by the Blackett family of Wallington until 1792, when it passed through an illegitimate daughter to the Beaumont family. The East and West Allendale mining areas were administered separately, the one from Allenheads, the other from Coalcleugh.

The Blackett/Beaumonts also leased most of the mines in Weardale belonging to the Bishop of Durham. Throughout the eighteenth and nineteenth centuries this family concern was the largest employer of labour in the region.[8] It worked all the mines in the western part of Weardale and in the tributary dale of Rookhope. A smaller group of mines to the south of Stanhope were leased jointly to the Blacketts and the Bacon family in the eighteenth century—the 'Partnership Mines' as they were called in the Blackett records. In 1791 the London Lead Company took over the leases of these mines and worked them for most of the nineteenth century.

The Derwent mines were worked in the eighteenth century by the London Lead Company, based at Blanchland. The company abandoned this area in 1806 and for a short time the mines were taken over by Easterby Hall & Co., then by the Derwent Mining Company, which built the new village of Hunstanworth from which to conduct its operations.

In the eighteenth century Teesdale was worked by numerous small groups about whom little has survived. The London Lead Company leased a few mines in the mid-eighteenth century; at the beginning of the nineteenth it greatly expanded its operations in the dale, taking over the smaller concerns. Middleton in Teesdale was the headquarters of the Lead Company's district agent.

The most important mines in the Tyne valley were Fallowfield and Settlingstones, both of which were worked in a desultory fashion throughout the eighteenth and nineteenth centuries. Evidence about how they were worked, the remuneration of the miners, etc., will be quoted later. The country around them, however, will not be examined, being outside the mining region proper. In the nineteenth century at least, these Tyne valley mines drew most of their labour force from Allendale.

This attempt to give a picture of the social and economic life of the

Plate I Two views of a semi-abandoned lead mine in the northern Pennines at the beginning of the twentieth century. The taller building on the right in the upper picture is a miners' shop. The lower picture shows a washing floor. *From T. Oliver*, Dangerous trades, *1902*

Plate II Miners climbing stemples.
From the 1864 report

lead mining region over a period of nearly two centuries is based on the analysis of many different types of source material. These are detailed in the bibliography, but some comment is necessary here to emphasise gaps in the evidence and unevenness in the surviving sources.

The most important manuscript sources have been the records of the lead mining concerns and ground landlords. Inevitably, these concentrate on issues of direct importance to the capitalist and landowner, and details of the lives of ordinary miners emerge only incidentally. As it has happened, only the records of the two largest mining concerns have survived in any volume, and their employees are the prime concern of this study.

The records of the Greenwich Hospital furnish much information about its policies as the major landowner on Alston Moor, and there are similar extensive records for the Blackett/Beaumont estates in Allendale. The vital period of parliamentary enclosure around the beginning of the nineteenth century is extensively documented in the collections of the land surveyors John and Thomas Bell.

The most detailed pictures of the whole region at given points in time are contained in the reports of the Children's Employment Commission of 1842 and the Commission on Conditions in Mines of 1864.[9] Of these, the 1842 report is vastly superior, quoting directly many statements made by miners and washer boys. One assistant commissioner in particular, Dr Mitchell, obviously made a genuine effort to describe conditions truthfully without bias towards one side or the other. The 1864 report, although it contains many useful statistics and appendices on particular aspects of mining life, is basically an assembly of statements made by the employers and their agents, cross-examined (not very critically) by members of the Commission. Taken together, the two reports give a comprehensive picture of lead mining life around the middle of the nineteenth century. No such sources exist for the eighteenth century, and here heavy reliance has had to be placed on the mining companies' records and the descriptions of such observers as Gabriel Jars and Sir Frederick Eden.

The more academic local historians have been generally less useful than other forms of local publication. Parson and White's *Directory* of Northumberland and Durham of 1827, for example, contains more relevant information than Surtees' *History of Durham* or Hodgson's *History of Northumberland*, which were published about the same time. A volume of essays by a Newcastle journalist, J. W. Allan, published in 1881, gives a better description of the mining industry in

B

decline than any more formal work. Critically gathered together, casual references in such works as these give a fairly detailed and, it is hoped, reliable picture of social life in the region at most times.

Lastly, one of the most important lead mining figures left a diary, covering fifty-six years of his life, in 167 volumes: Thomas Sopwith, land agent and surveyor on Alston Moor in the 1820's and 1830's, and chief mining agent of the Beaumont family from 1845 to 1871.[10] He was keenly interested in engineering progress, administrative efficiency and social reform. His technical and topographical works about the area are cited frequently, but his diary is uniquely important in its detail and concern with everyday mining life. His highly idiosyncratic views and comments are a welcome relief from the formal administrative records and government reports.

NOTES

[1] Adm. 66–105, p. 32.

[2] Greenwich Hospital report, 1823, p. 14.

[3] T. Sopwith, *Diary*, 9 July 1838; 7 January 1846.

[4] *Rating of mines*: select committee report, 1857, pp. 26–9.

[5] Following the practice of eighteenth- and nineteenth-century sources, the word 'miners' is used to mean (generally) all the workmen engaged in mining *and* (narrowly) those extracting the ore from the ground. The context should make the meaning obvious.

[6] M. Hughes, *Land, lead and coal*, 1963, pp. 43–69.

[7] Dr A. Raistrick has narrated the history of the London Lead Company's nation-wide activities in 'The London Lead Company' (Newcomen Society *Transactions*, vol. 14, 1933–34, pp. 119–62) and *Two centuries of industrial welfare*, 1938.

[8] In 1842 the Beaumonts employed just over 2,000 persons in Allendale and Weardale. Dr Mitchell (1842 report, p. 724) reckoned that this comprised two-fifths of the mining population. The London Lead Company, the second largest concern, employed about 1,500 persons at that time.

[9] The evidence for these reports was obviously gathered previous to their publication, but the convention has been followed throughout of quoting evidence from them as referring to their respective years of publication.

[10] Selections from the diary were published by B. W. Richardson in 1891. This single volume contains only a minute proportion of the original diary.

II

The lead mines

The geological nature of the ore field was directly responsible both for the nature of the mines and for the whole organisational structure of the industry, so different from that appertaining in the neighbouring coalfields.[1] The lead ore consists of galena mixed with various other minerals such as fluorite, barite, quartz and calcite (known to the miners as 'spar') and, near the surface, with iron hydroxides. It was introduced into faults and fissures in the carboniferous limestone by mineralising fluids acting in the remote past; the origin of these fluids is still a matter of controversy. The mineralised faults or fissures are known as veins; usually they are nearly vertical in limestones and sandstones, and the ore bodies vary from a few inches up to twenty feet wide. They may extend for some thousands of feet horizontally, but their vertical dimensions are measured in tens or at the most hundreds of feet only. Long stretches of the fissures are barren between the ore bodies. Workable ground seldom occurs in the shales, where the inclination of the fissures is lower than in limestone or sandstone. Associated with a few of the veins, horizontal replacement deposits occur in limestone—particularly, though not exclusively, the Great Limestone; they are known as flats and may extend as much as 300 ft from the vein.

The productive veins lie mainly in the east-north-east and west-north-west directions across the mining field, with lesser cross veins intersecting them. They are by no means homogeneous, either in their form or their contents. As the vein descends into the ground it may narrow to a point, or else 'hade'—slant from side to side—giving a zig-zag course. Wherever two veins intersect, one of them may be thrown off its course in a lateral direction, to reappear again several feet further on, on the opposite side of the main vein.

The contents of the vein vary according to the strata through which it passes. In the Plate beds (shale) the vein is filled with a soft clayey substance called 'donk' or 'dunk' by the miners. Where lead ore does occur, it does not fill the whole vein but is interspersed with ribs of spar, the whole being known as 'rider'. Sometimes the vein may consist wholly of spar; sometimes ore may appear only lower down. A vein of ore two inches thick was considered worth working in the nineteenth century.[2]

This complex and irregular system of veins made lead mining a far more uncertain business than coal mining. It provoked much geological thought and investigation by the miners themselves. Until the end of the eighteenth century the knowledge accumulated remained largely traditional, communicated orally, but in the nineteenth century a number of works were written by miners on the geology of the area. Westgarth Forster's *Treatise on a section of the strata, from Newcastle upon Tyne, to Cross Fell in Cumberland; with remarks on mineral veins in general* . . . (first edition, 1809) was the earliest and best of them. A third edition was published virtually unchanged as late as 1883. A later book, summing up in its title the central problem that faced all miners in approaching the ore field, was William Wallace's *Laws which regulate the deposition of lead ore in veins* (1861). As far as the general layout of the ore field was concerned, the courses of the main veins were known from an early date; maps showing them were produced, both in manuscript by the mining companies and privately printed and published. But in spite of the increase in geological knowledge, mining remained a risky and uncertain business—both for the entrepreneurs and, as will be seen later, for the workmen.

The method of working

When the presence of a vein was suspected, the initial investigation was by means of a process known as 'hushing'. A stream was dammed in an appropriate place and the accumulated water released when a sufficient quantity had built up. The flood of water swept away the surface soil and peat in its path, exposing the underlying rock. In the eighteenth century and earlier this technique was used as a crude method of extracting the surface ore; later, its prime use was in prospecting to see whether a place was worth mining. The surface proprietors of the land naturally did not like hushing, as it carried away alluvial soil and frequently caused floods miles below the site of the actual hush.

Once the presence of a vein had been established, the ore was extracted by means of a system of shafts and galleries, the scale and overall design of which progressed greatly during the eighteenth and nineteenth centuries, independently of the mechanical advances that occurred throughout the same period. At the beginning of the eighteenth century, access to the vein was mainly by shaft—the quickest and easiest approach. The cheapest and most primitive form of 'mine' was a simple pit, or a series of pits, along the course of the vein. In 1691 the *Journal of the lead mines* at Fallowfield recorded 'the disposal and letting the Shafts termed Bargains'. Galleries, or 'levels', to the outside world were rarely very long, and were constructed mainly for the purpose of drainage. 'They were made so as merely to admit the workmen, and were from 4 to 5 feet high, 2 feet wide near the top, and from 15 to 18 inches at the bottom. Though called *levels*, their inclination often varied with the strata, and they sometimes rose so much as the rapid ascent of 45°.'[3] The construction of long levels was expensive, as they entailed 'dead work'—work which did not produce ore—with no immediate reward, and often a long period of time before completion. The smaller companies and groups of adventurers could not afford the capital for extensive dead work, even if it would eventually result in far greater returns. Thus technical advances were made chiefly by the three richest and largest concerns—the Blackett/Beaumonts, the London Lead Company and the Greenwich Hospital. The latter, although not itself concerned with extracting ore, was prepared to give active help to its lessees on Alston Moor.

The advantages afforded by approaching a vein by means of a level rather than a shaft were threefold. First, as the mining field was in a mountainous area, it was possible to drive levels from the 'day' into or near the ore-bearing strata; this greatly eased the labour necessary to extract the ore and 'deads' from the mine to the surface. Second, a level at a sufficiently low altitude would drain all workings above it—and in doing so it would collect the water necessary to drive machinery for pumping out still lower parts of the mine. Third, these long levels might well strike a new vein in the course of construction.

The chronology of the introduction of long levels and the systematic planning of mines is uncertain. A nineteenth-century miner stated that 'it was not before the middle of the last century that the mines were first opened out and drained in a systematic way.'[4] This is probably fairly accurate; it was not, however, a sudden discovery but a gradual evolution. In 1737, for example, one mine lessee on Alston Moor wrote quite casually that if the water did not 'goe off there will be obliged to bring up a levell about 160 yards.' Another level, mentioned in 1738, was over 600 yards in length.[5] But these levels were intended for drainage and not for access to the mine. By 1765, however, there was a level at Coalcleugh containing a horse-drawn waggonway a mile in length.[6]

The longest of these long levels were intended to 'unwater' not just one but a whole series of mines. The Nent Force level was commenced by the Greenwich Hospital in 1776 to drain old mines and discover new veins along the valley of the Nent between Alston and Nenthead. It was driven forward intermittently until 1842, when it was over four miles long.[7] An equally ambitious plan was the Blackett level, driven up the East Allen by Sopwith and commenced in 1855. It reached a total length of nearly four-and-three-quarter miles.

By the nineteenth century the opening of a new mine was approached systematically—by the larger concerns at least; the smaller groups of adventurers on Alston Moor went on much as they always had. The chief agent of the London Lead Company described the procedure to the 1864 commissioners:

When we set on a level it is seldom for the purpose of covering only one object; we, generally speaking, know of the existence of a number of veins before us, and on our reaching the vein, if it be of sufficient magnitude and of a fair mineral character, then we rise to the surface, not only for ventilation, but for proving the character of the vein in all the strata between the level and the surface; we follow up the vein according to the course of the lode . . . Into . . . [the original] level we put a horse and waggons always. Suppose we find, in the course of our rise, in the different strata through which we have passed, ore-yielding indications, then we put on, from the rise, drifts in the course of the vein; and between such drift and the horse level, and also the surface, frequent communications are made for ventilation and other purposes. As each vein is reached, a like procedure takes place.[8]

The scale such mine systems could reach is illustrated by Burtree Pasture mine, in Weardale. In 1864 the horse or main level was two miles long. The end of it was 105 fathoms below the surface; below it another shaft went down a further 131 fathoms. In all, there were five levels below the horse level.[9]

The dimensions of the levels were also standardised. A bargain formula much repeated in the Beaumont records from the end of the eighteenth century onwards was that the miners should construct the levels '6 feet high upon the sleepers, and 3½ feet wide.' By 1864 the London Lead Company's horse levels had been standardised at 6½ ft × 4½ ft.[10] All levels rose slightly as they progressed into the hill to allow

water to flow out—some 4 ft in every 1,000 in the best Blackett/
Beaumont and London Lead Company levels.

In the early and mid-eighteenth century the levels and shafts were
generally very badly supported. Jars commented in 1765, after a visit
to Coalcleugh and to the London Lead Company's mine at Rampgill,
that although the rock required little support this little was rarely done
at all properly. 'Il y a . . . une négligence très-grande à cet égard: j'ai
passé dans plusieurs de ces galeries où les bois étoient pourris et cassés,
et menaçoient d'une danger évident.'[11] By the nineteenth century
things had altered for the better. A Greenwich Hospital report of 1822
commented that 'the walling of shafts, and the arching of the levels
was another great improvement, many of the old excavations having
fallen in, whereas now the mines are rendered accessible at any future
period.'[12] In 1864 the London Lead Company's agent said that 'We
always arch our levels with stone arches, as a matter of course, that is,
our horse levels. I may say that we universally arch our horse levels.'[13]
At Coalcleugh, however, although some of the horse level was arched
or supported with wood, for most of its length no supports of any kind
were used; the natural rock was considered sufficiently solid.[14] Shafts
were surrounded by dry stone walling, at least when they reached the
surface and where they passed through loose strata.

By the end of the eighteenth century, access to a mine was normally
by a straightforward walk along a level. Shafts were relatively short,
communicating between one level and another, and were rarely the
main route from the surface. In 1765 Jars had found no ladders at all
in the mines he visited in the area.[15] The shafts were ascended and
descended by means of 'stemples' [see plate II]. These were 'pieces of
wood placed at two opposite sides (of the shaft), 4 or 5 feet above each
other.'[16] A frequent payment note in the Blackett/Beaumont records
in the eighteenth century was 'for hewing stemples.' Sopwith quotes a
visitor to an Alston Moor mine in 1820 who had 'to scramble 30 feet
from one cross rafter to another, by hands and feet; an exertion of some
difficulty, for these cross sticks were in places so far distant as to
require all the active agility of youth to mount them.'[17] These
stemples were still in common use all over the mining field in the early
nineteenth century, but by 1864 ladders were far more common, save
in the smaller mines. No stemples were used at all in the London
Lead Company's mines by then, as it had been found that 'these
stemplings are very apt to rot at the ends that go into the rock, and
they might break almost without the knowledge of the parties that

they were so near the point of breaking.'[18] But access to most mines
was by means of a level. Miners were saved the wearisome climbs of
many hundreds of feet up and down ladderways which were usual in
Cornish mines.

At many mines where shafts were the chief means of access in 1864,
mechanical means were used to lower and raise the men. At Allenheads there were two cages in the Gin Hill shaft, each taking three
men at a time down the 77 fathoms to the foot.[19] In 1842 access here
had been by ladders from the surface, unless men walked down the
valley to the level mouth.[20] In the Derwent mines, which because of
the flatter nature of the surface and the depth of the vein had always
been worked mainly by shafts, the agent in 1872 noted that 'We
offered 2 Men and 2 Boys a place at Jeffries, but they refused to go
because of the long ladder-ways.'[21] All the other access shafts in the
Derwent area had mechanical hoists by this date.

As for the actual ore workings, these were

of almost every conceivable size and shape, from 2 to 20 feet wide, and 6 to
60 feet high . . . . The ore is worked by headings, roofs and soles; when a
roof working gets to a height beyond reach, a timber floor or 'bonding' is
fixed, from which the miners continue their operations. When the ore is
found in 'flats' the height is from 3 to 6 feet, and up to 60 or 70 feet wide on
each side of the vein, but the width is exceedingly irregular.[22]

As the overall design of the lead mines improved, so did mechanical
methods of assisting the miners in their work. But the basic step of
mechanising the individual miner by providing him with a power-
driven rock drill was not taken. Experiments were carried out by
Sopwith for the Beaumont concern, and by the Derwent Mining
Company, but they did not progress beyond the experimental phase.[23]
The miner of 1870 relied on tools worked by his own muscles as much
as did his predecessor a century and a half earlier. Gunpowder was the
only mechanical assistance either received. The advances in mechanisation were in the subsidiary fields of underground haulage, drainage
and ventilation.

The disposal of the excavated material—both ore and deads—was
one of the biggest problems. The ore, of course, had to be brought to
the surface. The deads need not be carried out—if somewhere suitable
could be found for disposing of them in worked-out portions of the
mine. But it was good mine planning to forbid this save in places never
likely to be either worked again or used for the passage of men.

Unfortunately this was something that could never be predicted with certainty. Improvements in washing practice at the beginning of the nineteenth century meant that veins could be worked which had been left as uneconomic in the eighteenth. Coalcleugh low level was described by a new agent in 1813 as 'in a very bad state, being ... entirely filled with Deads both East and West from the Engine.'[24] This was greatly hampering new exploration as well as impeding ventilation. In a well planned mine the great majority of deads, as well as the ore, had somehow to be brought to the surface.

Before long access levels were introduced, all the ore and deads had to be lifted from the mine by hand. They were dragged in barrows along levels and loaded into pails called 'kibbles' to be drawn up the main shaft by a winch or, in some cases, a horse-driven gin. This operation, known as kibbling, was called by one eighteenth-century writer 'the most laborious part of manual labour' in the mines.[25] In small mines, it continued to be performed in this way right through the nineteenth century. In large, deep mines this primitive form of extraction took up as much, or more, time as the actual driving of the levels. In the Coalcleugh low level example quoted above, the agent reported that 'the great expence of drawing Deads from such a deep field prevents many Men from attempting to make a search to raise Ore in places that I am of opinion would produce Ore, had they been in a good situation'.

The introduction of long levels saved a great proportion of the labour. Within these levels were laid railway tracks along which waggons ran to the open air. The earliest of these waggonways was installed at Coalcleugh in the period 1755–57.[26] This is apparent from the accounts of the mine, recording men 'laying waggon rails' and 'Trailing them up the Level'.[27] Jars visited Coalcleugh in 1765 and wrote:

Lorsque les minérais et les déblais ont été élevés par plusieurs puits des différentes profondeurs, soit à l'aide des treuils, soit par une machine à moulettes établie dans la mine, jusqu'au niveau de la galerie supérieure d'écoulement, elles sont extraites au jour par cette même galerie, dans laquelle on a pratiqué un chemin semblable aux nouvelles routes, dont on fait usage à Neucastle pour voiturer le charbon, et dans des mêmes chariots, avec cette différence, qu'au lieu de 4 pieds entre les deux pieces de bois sur lesquelles roulent lesdits chariots, il n'y en a que deux, et que ceux-ci sont plus petits et sur-tout trésbas; un seul cheval en conduit deux pleins de matiéres. La galerie est prolongée fort avant dans le silon, même jusqu'aux limites des deux exploitations;

et comme la compagnie de Londres n'a pas dans sa mine [Rampgill, Nent-head] la même aisance pour le transport de ses déblais, elle paie à l'autre compagnie pour le passage, une somme fixée par chaque chariot.[28]

By the beginning of the nineteenth century, all the major mines had horse, or 'galloway', levels. In a well designed mine they would connect directly with the washing floors. Cast iron rails replaced wooden ones in the period 1805–20,[29] but the waggons remained much as Jars described them—2 ft 6 in wide, 4 ft high, and 6 ft long, according to Sopwith in 1833.[30] The Nent Force level was unique in the area, being conceived as an underground canal. Some two and a half miles of the level were constructed in this way. It was then necessary to rise about 200 ft, so the remaining two and a half miles of tunnel were conventional.

Drainage

The other mechanical advance which made the lot of the lead miner easier was in the field of drainage. The engines originally developed to pump out the lower parts of mines became so powerful and sophisticated in the later nineteenth century that there was a surplus of power available to drive other machinery as well. The long drainage levels were very expensive to drive, but once it had been done they cost little to maintain. In 1778 a Greenwich Hospital agent criticised 'a practice which prevails very much to the detriment of the Lessees in Alstonmoor which is that of pumping Water instead of driving up Levels.' He instanced one company which 'in the price per Bing and expences of pumping, making convenience for Air and washing paid above twice the sum for raising Ore . . . than it would have cost had the Level been up.'[31] Pumping engines were costly and required constant maintenance.

But some form of pumping was always necessary in the very deep parts which were too low for a drainage level to be practicable. In some of the Derwent mines, and in those north of the Tyne at Stonecroft and Fallowfield, the terrain of the surrounding country and the depth of the veins prevented long levels being employed at all as the primary means of drainage. Here pumping was as essential as it was in the Cornish mines, where many galleries ran below sea level.

The most primitive method of removing water consisted simply in lifting it in buckets by hand. It remained the only way of draining isolated deep workings below the level of the pumping system. Soon after the middle of the eighteenth century, power-driven pumps were in common use to drain the greater part of the mine systems below the main drainage levels.

In Cornwall from the beginning of the eighteenth century the pumping engines were steam engines, fired mainly by coal. Despite the proximity of the coalfields, steam engines were unusual in the northern Pennines. The reason was that the topography was ideally suited to the use of water power. Steam engines were used in only two districts: Derwent, and the small mines north of the Tyne. In both cases, communications with the coalfields were better than in the case of the mines further west and south. Fallowfield had a steam engine in operation by 1773,[32] and coal-fired engines continued in use here and at the neighbouring mine of Stonecroft throughout the nineteenth century. In the Derwent mines, steam was twice replaced by water power. According to the first edition of Mackenzie's *Historical and descriptive view of . . . Northumberland*, published in 1811, the new lessees of the mines had replaced the water engines with steam engines in 1807, as the London Lead Company, the former lessee, had 'suffered the engines and gins to go to ruin'.[33] But in the second edition of Mackenzie's work, published in 1825, 'Water engines have again superceded the use of steam engines in the Lead Mines as, from great improvements made in water-wheels and water-pressure engines, they have been found to be nearly as effective, and far less expensive than steam engines'.[34] By 1842, however, there were three steam engines once more in operation in the Derwent area, working alongside the water engines. But water power finally triumphed and the last steam engine was replaced by a water pressure engine in 1872.[35]

In the greater part of the mining field, water power was never challenged. Particularly in the Blackett/Beaumont areas, hydraulic engineering developed to a very high degree of efficiency. It was at Coalcleugh in the early 1760's that the water pressure engine was invented, by William Westgarth, the resident agent. Jars described its principle when he visited the mine in 1765. The lowest parts of the mine were below the drainage level:

mais pour élever les eaux d'une plus grande profondeur, on est dans l'intention de profiter de celles qui passent en trés-grande quantité dans la galerie supér-ieure, pour faire mouvoir une machine d'une nouvelle construction à laquelle celle à feu a donné lieu. On se propose d'avoir un cylindre semblable, mais au lieu de la colonne d'air de toute la hauteur de l'atmosphere, on veut conduire sur le piston une trés-grande colonne d'eau.[36]

During the later eighteenth century and the nineteenth, reservoirs were constructed all over the region to harness water power. All the streams were made use of, some many times over. A somewhat florid description of the East Allen by a guide-book writer in 1859 runs:

From a reservoir on the hill-side, 180 feet above Allenheads, the water descends, impels the two hydraulic engines, does all the work of the washing floors, drives the lathes, the saws, the machinery in the machine-shops, keeps four wheels going deep underground, which pump water from the deepest levels of the mine (and drive haulage machinery) by an adit, it flows into the Allen about a mile to the north. Not long, however, does it run at liberty, for intercepted by a race, it flows along that for two miles to Breckon Hill—the first shaft on the Blackett Level—there turns two wheels that force water into the accumulator, and is discharged again into the Allen.

Lower down, the water worked many more wheels, including that of the Allen smelt mill.[37]

The miners' work

From the limited function of drainage in the 1760's, water power developed into a source of energy for virtually all mechanical operations on the mining field. But although it made a great difference to the productivity of mining operations—mines could be sunk to a greater depth, the ore and deads could be extracted with greater ease and despatch, the washing operation became far more efficient—it made remarkably little difference to the major part of the labour force, the actual miners. Technical advances meant that they were able to spend a greater part of their time in tunnelling and getting ore in the mid-nineteenth than they had been able to do in the mid-eighteenth century; veins could be worked which would have been unprofitable before. But the methods they used remained almost exactly the same.

The organisation and control of the partnerships of miners engaged in working each mine will be considered in chapter III. Here we are concerned with the jobs undertaken by the ordinary miners, and the techniques they used. When the presence of lead ore was suspected below the surface at a new place, the men were allowed to approach it in their own way. A 1746 bargain from Weardale runs:

Let to Joseph Featherstone, 4 ptnrs. a bargain to gett Oar by Hushing on the Float where Arthur Watson wrought last at 18/- per Bing . . . and Arthur Watson to have Liberty to sink where he pleases, not to interfere with their Hush.[38]

In 1807, also in Weardale, the initial approach was as vague as it had been sixty years before:

Two little boys had found some small pieces of Ore in the Sides of the Old Hush gutter, and in working further found the quantity of Ore increase, which encouraged some workmen to take a Bargain to raise Ore the last quarter who have made very good wages . . . at 50/– per Bing without anything for dead work.[39]

The construction of a new mine, or an extension of an old one— 'dead work' as against actual ore getting—was undertaken far more systematically. Men were instructed to sink a shaft, or a sump (a shaft not reaching the surface), or to drive a level, to and from given places:

Lett to Geo. Furnace & 7 ptnrs. a Bargain to Drive a Level from Ireshope Burn to the Vein they wrought in last and to Drive it immediately thro' the Sump foot at £60 per Sump and they're oblig'd to make a conduit from the burn side to the Bank foot half a yard high and a foot wide and to flagg it in the bottom.[40]

When it was the main or horse level under construction, the men who drove the level were also expected to lay the rails:

Let Jacob Lowe & 7 Partners a Bargain to drive their Level forehead 20 Fathoms more west, at an exact true level, sufficient large Drift, and lay the Vogue Rails at 48/– per Fathom.[41]

The actual driving of a level or the sinking of a shaft was done by hammer, pick and shovel where the ground was fairly soft or broken up, and by blasting with gunpowder as a preliminary where the rock was solid.

The vein of ore was preferably approached from below, and work proceeded upwards rather than downwards:

It is easier to work upwards than to work downwards, because in working upwards all the dust and broken pieces fall down, whereas in working downwards they accumulate at the bottom, and it is troublesome to remove them. The miners in their upward work make a small landing-place, and go from one stage to another, so that they may be able to place ladders or pieces of wood from side to side, and be afterwards able to climb up, and have halting places at short distances all the way.[42]

This was known as a 'rise'; when it was necessary to work downwards, access was by means of a sump. At the bottom of the rise, or top of the sump, the level was widened to allow work to be put into waggons more

conveniently. After each length from the rise had been worked out, it
was filled with the deads from the length above, leaving sufficient
room for access from the level. The rises might go up several fathoms,
but if the vein were sufficiently rich it was economic to drive a new
level at a higher altitude before the rise became too high for convenient
working.

Different geological conditions called for different methods of
approach to the vein. 'The most difficult part of mining', according to
Forster, was getting ore from 'soft veins', where the vein was surrounded by fragmented rock and had swollen 'out to an enormous
width.' Here the miner had to surround himself with square timbers
as he proceeded, whether on the level, upwards, or downwards:

The miners stand within this square timber, where they work, and still set
more timber before them, as they can make room for it. This is expensive,
troublesome, and dangerous, if they had not got good skill in setting the
timber, for the soil is generally quite soft and loose, and . . . the whole will
frequently rush down with violence before or in front of, their timber . . .
But it frequently happens, that the ore is so plentiful and good, in these veins,
as abundantly to compensate for all this trouble and expense.[43]

Another peculiar form of ore working was in a flat vein where the ore
had replaced a flat post or bed in the limestone. Many men could work
together in a 'flat', and the workings could 'frequently run to a great
extent, and the roof being supported by substantial posts of timber, the
whole, especially when dimly lighted, presents a receding vista, which
reminds the spectator of the aisles and pillars of a cathedral.'[44]

A frequent form of entry in the Blackett/Beaumont bargain books
was for miners to get ore 'in the roofs and sides which is left by the old
man.' Old workings were constantly re-worked by successive generations of miners as improvements in mine planning allowed a better
circulation of air to places where the lack of oxygen had formerly been
almost lethal; and progressive advances in the efficiency of the washing
machinery caused veins that had been abandoned as uneconomic to be
worth reviving. Here, the miners' work was much as it was elsewhere
once the vein had been reached, but complicated by the huge piles of
deads that jammed abandoned levels and sumps, acting as a barrier
both to the circulation of air and to the extraction of fresh ore and deads.

When the vein was soft the ore was extracted by pick and hammer.
More usually it was necessary to have recourse to blasting. A hole was
first driven into the rock by means of a steel-pointed iron bar, about

2 ft long, known as a 'jumper'. This was held by one man who slowly rotated it while another hammered it into the rock. When the hole had been bored, a paper cartridge of gunpowder was inserted. A thin iron rod known as a 'pricker' was then driven in through the cartridge to the very end of the hole. Around the pricker were placed small pieces of shale or clay, pushed in by means of a 'driver'. Finally, the pricker was withdrawn and replaced by a paper fuse. All but one of the men concerned then withdrew to a safe distance.

The man who has to fire off then lights the match, and runs off as fast as he can, and presently the shot goes off with much noise, smoke, and dust. The men return and find a chasm made in the rock, and with hammers and picks they strike upon every projecting piece of rock, and bring it down. The chamber where they work is now full of smoke, and every additional shot fired off makes the place worse and worse as they continue their work throughout the day.[45]

The method of blasting outlined above is given in virtually identical terms by Sopwith, writing in 1833, and by the 1842 and 1864 commissions' reports. Thomas Crawhall's article of 1822, on the discovery of old blasting implements in the Allenheads lead mine, makes it clear that the process and the implements used in it had been established for at least a century before that.[46] Few improvements were made during the nineteenth century. Between 1842 and 1864 the iron pricker was generally replaced by one made of copper, which did not tend to spark in contact with rock. By 1870 gun cotton had replaced gunpowder in many of the Beaumont mines.[47] A partnership might fire off 'six or eight shots a day'.[48] A great deal of dust would be created, not only by the actual explosions but also by the constant boring and hammering, and some of it inevitably found its way into the workmen's lungs.

When the vein had been opened out, timber was put up as roof supports where the men considered it necessary. A Beaumont agent told the 1864 commission that the workmen were given 'the choice of what timber they think proper, in the woodyard'. This was supplied to them free on request, although 'sometimes we press it on them, and think that they ought to have more'.[49] In many places, however, the veins the ore workings were following ran within solid rock that needed hardly any support.

The extraction of ore from the vein was only the first stage. It had then to be raised to the surface, accompanied by considerable quantities of deads of no value but for which there was no room within the mine. As will be seen in chapter III, the responsibility for dealing with deads normally lay with the partnership who originally cut them. When engaged in getting ore, the partnerships' earnings depended solely on the amount of ore they raised: no extra payment was made for the laborious task of extracting deads. Left alone, the workmen tended to block useful space with deads as well as worked-out space. Unless the practice was strictly controlled by the agents, it hampered future explorations severely, as happened at Coalcleugh in 1813. Hence the necessity of carrying large quantities of deads to the surface.

Even allowing for the revolution effected by the use of rails and the introduction of the horse level, an enormous quantity of the work had to be transported by manual labour. The Coalcleugh agent, reporting on a partnership that had raised more than 200 bings in a quarter in 1807, commented:

It is almost incredible to believe the 8 men should go through such a quantity of Work, considering the disconvenience attending their getting it to Bank, having first a Sump to draw it up about six fathoms, and carry it back to another Sump, and then draw it up about 20 fathoms. It is then taken to the Whimsey on a Waggon by a Horse about 260 fathoms and drawn up the Whimsey, and taken by Horses down the Level to the surface, from which you will easily see the Expense which must attend the Banking of every Shift of Bouse Ore.[50]

A London Lead Company miner complained in 1842 that the mines could be better designed to cut down the hard work necessary.

The levels often might be made so as to allow the horse and waggon to go further on, nearer to where the men work, and thereby save the laborious and fatiguing work of carrying the ore in barrows to the horse; the miners have to do this work themselves and this prevents getting so much done of what is properly the miner's work, and keeps down their gains . . . . This severe laborious work injures men much more than the regular work.[51]

There was one other important task that all miners occasionally had to face. This was the problem of dealing with water. By the end of the eighteenth century most large mines had long drainage levels, together with water pressure engines to drain the parts of the mine below the drainage level. The pumps, however, sometimes failed. In 1786 a disastrous flood occurred at Allenheads, when the river changed its course and went through an old shaft into the mine. It was nearly three months before the mine could be worked properly again.[52]

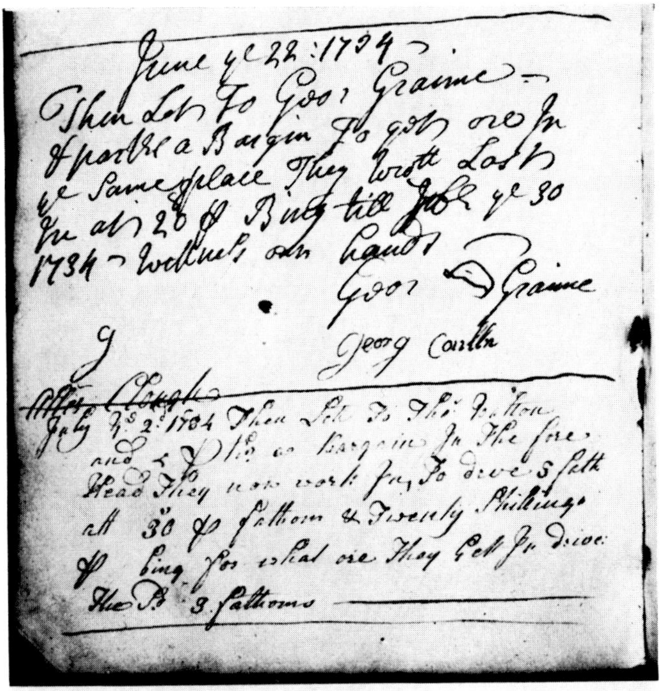

June ye 22: 1734
Then Lot To Geo: Graime
& partners a Bargin To get ore In
ye Same place They wrott Last
In at 20/ ⅌ Bing till July ye 30
1734 – Witness our hands
Geo: Graime
Georg Carth

Alex Clough
July ye 2: 1734 Then Let To Tho: Walton
and Partners a bargain In The fore
Head They now work In, To drive 5 fath
at 30/ ⅌ fathom & Twenty Shillings
⅌ Bing for what ore They Get In drive:
the Pt 3 fathoms

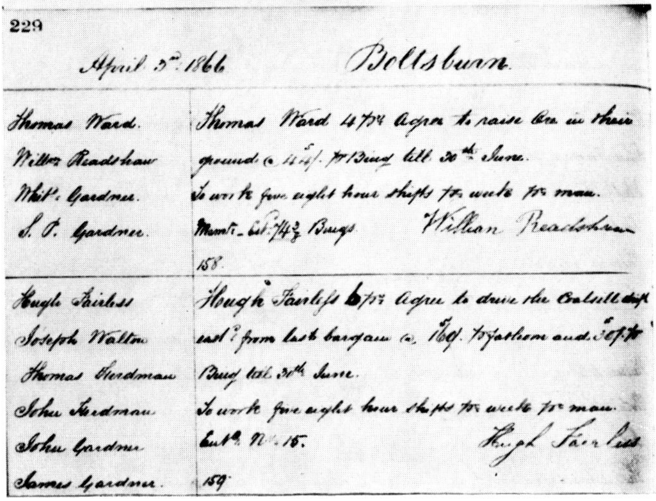

229

April 5th 1866 Boltsburn

Thomas Ward.	Thomas Ward & Prs Agree to raise Ore in their
Willm Readshaw	ground @ 14/- ⅌ Bing till 30th June.
Whit. Gardner.	To work five eight hour shifts ⅌ week ⅌ man.
J. O. Gardner.	Ment. Est. 74½ Bings William Readshaw
	158.
Hugh Fairless	Hugh Fairless & Prs Agree to drive the Coalcleugh drift
Joseph Walton	east from last bargain @ 16/- ⅌ fathom and 30/- ⅌
Thomas Herdman	Bing till 30th June.
John Herdman	To work five eight hour shifts ⅌ week ⅌ man.
John Gardner	Est. No. 15. Hugh Fairless
James Gardner.	159

Plate III Entries from the Blackett/Beaumont bargain books, June–July 1734 and April 1866

Plate IV   Equipment used for washing lead ore at the end of the eighteenth century. *From the Mulcaster MS, Newcastle Literary and Philosophical Society*

Working could also be halted by a drought. At Crowlah mine in Weardale in 1818, 'We have got little done the last quarter for want of a supply of day Water to work the Engine. The men have been off work for seven weeks.' In cases like these drainage by hand supplemented the pumps.

Individual workings, too, were frequently below the range of the pumps. Partnerships earned more money for working there, either in the form of direct payment for keeping the workings drained, or by receiving higher sums per bing of ore. Two bargains at Breckonsike mine in Weardale in 1762 show the distinction. One partnership, in addition to the main bargain, was 'to draw the water at the 2 high Sumps ... at 5/6 each a Week and to find their own Candles.' Another, receiving a higher price per bing, was 'to keep the Water out of their low bottom the whole Quarter at their own charge.'[53] Sometimes the work was too much for the partnership concerned, either because the influx of water was too strong, so that 'the water quit them out'[54] or because it became economically impossible to continue the bargain.[55] In normal circumstances only one or two partnerships in a mine had to drain water by hand. The mine owners paid men larger sums to get them to work in such situations. If more than a few partnerships were working below the range of the existing pump, it became 'as much expense [to the owner] drawing the Water as would pay the expense of Engine.'[56]

The work of drawing the water with kibbles was irksome and unpleasant. But the miners did not mind a moderate amount of water. As one agent told the 1864 commission, 'I remember some who worked in a dry mine and died sooner than they would have done had they been working in a wet mine, so far as I was capable of judging.'[57] This observation was probably quite true, as the dampness absorbed the dust from drilling and blasting, and lessened the amount entering the miners' lungs.

There was very little specialisation among the lead miners. They were prepared, and indeed expected, to carry out all types of work underground, though boys were employed to drive the waggons removing ore and deads from the mine. They were also employed to pump the 'air machines' blowing air along tubes into remote workings. Most of the relatively few specialist jobs for adult workers were on the surface. Masons, smiths, millwrights and joiners figure in the Blackett/ Beaumont accounts from 1730 to 1870. Their jobs were strictly subsidiary to the partnerships raising ore. 'Woodmen' were responsible

c

for the maintenance of the main levels and shafts. In 1860 a Weardale partnership of woodmen was instructed to 'Keep in good repair all those Mines or portions of mines specified ... including Horse Levels, cross cuts, rises, Wagon drifts, top or Air drifts from Horse Level Mouth to the Forehead and from the Wagon drift to Blakes Laws Ladder'd way top, including all railways.'[58]

Underground, even jobs like the maintenance of pumping machinery were left to the ordinary partnerships. One partnership in 1760, in addition to a bargain to raise ore, was 'to assist Thos. Bradshaw at Leathering the Engine and all things relating thereto all the time of their Bargain at their own charge.'[59] The Blackett/Beaumont accounts show no notable change in this policy of non-specialisation of workmen between 1720 and 1880. There was, perhaps, a proportionately greater number of masons, millwrights, etc., and there were also a few new jobs—or at least new job titles—such as 'engineers' and 'apprentice engineers' in the later period, but it was a very slight increase. At all periods, if a special task had to be done, an ordinary partnership would be asked to do it on a normal quarterly contract basis.

This brief account of the work and conditions of lead miners has been based on the evidence available, and the evidence leaves many factors unrecorded. The sanitation arrangements within the mine (if any) are not mentioned in the reports of either the 1842 or the 1864 commissions, though one of the witnesses examined in 1842 does remark on the presence of rats underground. As the normal access route to most mines was by way of the main drainage level, many miners started and ended the day with wet feet, even if their actual place of work was dry. These may be considered the minor discomforts of a lead miner's employment. The two commissions did, however, supply evidence about more acute dangers to life and health.

Compared with coal mining, fatal accidents were comparatively rare. The main reason was the virtual absence of explosive gases from the lead-bearing strata. There were only twenty-five fatalities among the employees of the London Lead Company between 1800 and 1875.[60] There were five deaths in the Beaumont mines between 1845 and 1862, and the most serious accident in the Derwent mines between 1855 and 1862 was the loss of one eye by a workman.[61]

An analysis of accidents in the 1842 report isolates seven causes.[62] Minor explosions due to explosive gases were not completely unknown, but scorching was the only injury they caused. Carbon dioxide— 'choke damp'—sometimes disabled men temporarily. Minor accidents were caused by falls down shafts and crushing by waggons. But the greatest sources of danger were collapsing roofs in the levels, and premature explosions while blasting. Six men died in Weardale in 1838 when the roof collapsed on them. Premature explosions were caused both by 'gross carelessness on the part of the man himself' and by deficient equipment supplied by the master. The witnesses before the 1864 commission were all questioned in considerable detail about accidents caused by blasting. The two chief causes appeared to have been deficient fuses, and the use of iron prickers to clear the way for the insertion of the fuse. The fuse was made basically of paper, and prepared by the men themselves. If badly done it could cause the powder to explode prematurely. The London Lead Company had offered to supply its employees with patent safety fuses, but after experiments most of them declined—'They do not like the smell of it at all.'[63] The iron prickers sometimes caused sparks to fly from the rock, thus igniting the powder while the men were standing near. Copper prickers obviated this risk.

Ventilation

The greatest threat to a lead miner's health lay in the atmosphere within the mine. Choke damp was frequently met with. It occurred in pockets in the rock strata and filtered slowly out into the workings. After a strike in Weardale in 1818 the agent noted that 'foul air' had set 'into a part of the Workings while they were off work, which continued a month before it could be extricated to get regularly forward'.[64] It could escape quite suddenly. During the driving of the Nent Force level in the early nineteenth century a bell was rung if gas suddenly appeared at the forehead, and the miners would exit at speed.[65]

Choke damp was fairly easily detected. Its poisonous nature meant that men would not attempt to work in it at all. Less obviously deleterious to health, and therefore more dangerous, was the air in the extremities of mines, where scarcely any natural circulation took place. One miner told Dr Mitchell in 1842 that he knew of mines 'in which there were levels 90 fathoms below the first level, and no other means of getting [air] but from the first level. The deeper men work, and the further distance they penetrate, the worse the air is.'[66] In this close atmosphere the miners were constantly blasting and drilling. Gunpowder fumes and minute particles of rock and ore floated in the air. After blasting 'we smoke our pipe till it gets off',[67] but time was money. They went back to work as soon as they could see. The seven or eight 'shots' a day that each partnership customarily fired meant that by the end of the shift the men were working in a continual haze. However, unlike the custom in Cornwall, the mine was normally worked only eight out of the twenty-four hours in a day. This did allow some clearance of dust by the start of each new shift.

The temperature within the mines was fairly constant summer and winter, and had little effect on ventilation. What did affect the atmosphere was barometric pressure. 'The lower the barometer the worse the mine is for air and so the miners know before they leave the mine whether it is raining or likely to do so when they reach the "day".'[68] In the summer, 'this being the worst season for air in the Mines, the Men cannot work more than half the usual time' in places where natural ventilation was bad and no artificial means of promoting circulation of the air were employed.[69]

The mines were ventilated by 'natural' and 'artificial' means. Natural ventilation meant that the mine was specially designed to procure a natural flow of air through all the levels and shafts. Artificial ventilation was the supply of air pumped through tubes to the foreheads where a natural flow was impossible.

In the eighteenth century many small mines were scarcely more than outsize pits, ending when air became too scarce. The larger mines were designed in a very haphazard fashion. In the Blackett/Beaumont records there are frequent bargains such as 'Thos. Waugh . . . to drive west in the vein at 17/6 per fathom. Wanted Air: let them to sink into a low drift at the same price.'[70] In other words, if the miners needed more ventilation they were allowed to make a shaft to the surface or to another level to promote circulation as they thought best. Jars wrote that there was no understanding of the principles of natural ventilation: 'Si les directeurs entendoient bien la théorie de la circulation de l'air dans les mines, il seroit facile de chaffer ce mauvais air par les tuyaux de communication d'un ouvrage à l'autre.'[71]

By the nineteenth century the problem was approached more scientifically. In the lead mines, with their rambling layout and numerous exits, it was difficult to use the system of planned air flow by means of trap doors and furnaces that the more regular coal workings permitted. But certain fundamental atmospheric laws could be ascertained and made use of. It was noted that the shaft at the highest part of the mine would generally draw the warm air from other parts to escape into the colder atmosphere outside. This air would be replaced by a flow into the levels and the shafts at lower altitudes. Forster wrote in 1821 that 'Where it is practicable to obtain a Gallery, which shall lead from the bottom of the Shaft to the day or open air, a current is easily established by this simple artifice; but if this is not possible, a second Shaft is sunk, at the extremity of the Gallery, opposite to the first, and if it opens at a different Level from the other, a circulation and consequent renewal of air immediately takes place.'[72]

By 1864 the larger mining companies were prepared to undertake extensive dead work to improve ventilation. In the Derwent mines, sumps connecting tunnels were driven so as to bring proper circulation of air as near to the men driving a new level or getting ore at the forehead as possible. One Derwent Company sub-agent, who had been a miner for over fifty years, told the 1864 commission that this was a new policy introduced by a new owner of the mines. Previously, smoke from a single 'shot' would hang around for hours; in 1864 the bulk would disappear in ten minutes.[73] In the Rodderup Fell mines 'about four fathoms above the level we have an air drift which we drive up, so as always to keep up with the level.' The two would be connected by a sump about every fifteen fathoms.[74] Both the London Lead Company and the Beaumont concern used trap doors to increase air flow in the places where it was most needed. The latter had opened two furnaces, one at Allenheads, the other at Burtree Pasture in Weardale. These were designed to heat the air at the up-cast shaft, thus stimulating the draught. It was found that they greatly improved ventilation even in places as much as a mile away from the furnace. 'In one particular part of the mine prior to the furnace at Allenheads being lighted ... the men in dull damp weather were frequently prevented from working by a want of ventilation, and since that they have not lost any work, and speak very favourably of the effects of the furnace.'[75]

In the blind extremities of levels and shafts, the fore-heads, it was very difficult to get a flow of air by natural means. Here it was often necessary to pump air along tubes for the men to be able to work. In the eighteenth century this form of ventilation was used extensively, as being cheaper than the dead work necessary for obtaining a natural flow. By the nineteenth century the larger mining concerns did apparently limit the use of 'artificial' ventilation to the places it was virtually impossible to ventilate naturally.

The most powerful form of artificial ventilation used in the eighteenth century was the water blast. A member of the Newcastle Literary and Philosophical Society, visiting Allenheads in 1793, described the system subsequently at a meeting of the society. 'It consists merely of a tub, with a perpendicular pipe, with many holes drilled in its sides, fixed in the top, and a bent pipe turned up from the bottom, making, with the tub, an inverted syphon.' This was placed under a fall of water down a shaft within the mine. 'The water, in its descent, forces down along with it the air in the pipe, which is immediately supplied by the holes bored in its sides at different distances.' When the air trapped in the water fell into the tub it had 'no way of escape but along a horizontal square pipe or box inserted into the side of the tub near the top, which, being carried along the roof of the level drift, supplies the miners with air to a vast distance.' It was found that the stream of air from the end of the pipe, 400 fathoms away, was 'so strong as nearly to blow out a candle'.[76]

This form of ventilation was in use at least as early as the 1730's in the Blackett mines, as numerous references to its installation in particular places show. It was still in common use in 1864, sometimes with a miniature pressure engine to give more force. In the eighteenth century the piping was all made of wood, which easily rotted at the joints, allowing air to escape. Iron and lead pipes were introduced early in the nineteenth century.[77] In 1864 it was the normal policy of both the London Lead Company and the Beaumont concern to install a water blast in every new long level as it was being driven.[78]

The great disadvantage of the water blast system was that good air came in *via* the tube; it was the bad air that was forced out along the level. 'The great danger then is of the foul air accumulating behind and clogging up . . . I have known instances of that kind where the men have had to come through some portion of the drift where they could not carry a lighted candle,' the Allenheads agent told the 1864 commission.[79] In the Blackett level the normal system had been reversed. The hydraulic engine at the water blast sucked air from the fore-head, causing the fresh air to approach along the level.

For more local use in individual ore workings the company would pay for the installation of piping through which air would be pumped by a boy in the main level. A frequent addition to bargain agreements in the Blackett/Beaumont accounts was such a phrase as 'and they are to have 1/11d per Week for finding lads to blow the bellows.'[80]

These were the systems in use to ventilate the mines. How effective were they? In a later chapter it will be shown that most lead miners died at an abnormally early age, owing to respiratory diseases brought about by the air they breathed in the mines. The dust and smoke-laden air was not immediately lethal, like choke damp; when choke damp leaked into a level the miners left at once. The contaminated air they had to put up with, or lose their occupation altogether. Assistant commissioner Mitchell wrote in 1842 that 'in most mines there are not two levels communicating with the open air . . . Where nature does not interpose a physical impossibility, there is another that is equally powerful—the dread of expense. The sum required to sink a shaft or drive a level may be so great that the mine is not worth it. The proprietor would rather discontinue working it than submit to the burthen; and the men, young persons, and boys, having no other means of existence, are eager to be allowed to work at the mine such as it is.'[81]

The evidence supplied by the Blackett/Beaumont bargain books and the cross-examinations of witnesses in 1842 and 1864 does not expose deliberate cruelty by the masters and agents, but a realisation by both master and men of the need for work to be continued in bad air. An agent's report on a Weardale mine in 1809, for example, stated that he had 'been under the necessity of advancing the price this quarter as the sinking to give the Air to the Mine is not yet completed.'[82] A Weardale bargain of 1747 gave a partnership 'liberty to work in the old yards when they want air in their working, not to neglect their working when the air will permit.'[83] Shorter hours were permitted for men working in bad places, and during damp days in the summer months many places would not be worked at all. But the miners were prepared to endure very bad conditions for the sake of extra money. One miner said in 1842, 'It is not proper to stop after the candle goes out'. But 'sometimes we must hang the Candle half perpendicular, that the grease may flow down to the flame of the candle, and burn in spite of the foul air. It is seldom necessary to do this, but perhaps it may be a place in which this must be done.'[84] Air blowing machines were supplied at the request of the men, provided the agents considered it necessary.[85] Men were allowed to leave their workings, according to a Beaumont agent in 1864, 'at their own discretion.'[86]

There was improvement in ventilation practices during the eighteenth and nineteenth centuries. Statistically, however, there was no improvement in the mortality rate of miners in 1864 compared with 1842. The men and agents asserted, in both years, that conditions had improved substantially. Dr Peacock told the 1864 commission that

'The reports of the older men, and of the agents and medical officers, and indeed of almost everyone connected with the mines, show that formerly the state of the mines as to ventilation was very much more defective than at present', but in spite of the improvements 'I was repeatedly told by miners now at work "that all miners have to breathe bad air, more or less", or that in "all mines there are places where the air is bad".'[87] A Beaumont agent thought that productivity had gone up ten per cent compared with a period forty years before 1864 as a direct result of improvements in ventilation.[88]

If a miner of the 1750's had returned a century later, he would have found the mines better drained and ventilated, his own access and the removal of ore and deads greatly eased by the construction of horse levels, and more machinery in use for pumping air and water. When he came to the fore-head, however, he would have found the same method of work and even the same tools as he was used to. It is probable that had the mines remained open until the later years of the nineteenth century greater changes would have taken place, such as the wholesale introduction of the power-driven rock drill. As it was, most mines closed down whilst these improvements were still in the experimental stage, and to the last the miners worked by hand as they had done for the previous two hundred years.

NOTES

[1] For a comprehensive account of the geology of the area, see K. C. Dunham, *Geology of the northern Pennine orefield*, vol. 1, 1948.

[2] T. Sopwith, *Alston Moor*, 1833, p. 141.

[3] Sopwith, 1833, p. 18.

[4] W. Wallace, *Alston Moor*, 1890, p. 140.

[5] Adm. 79/35–4; Adm. 66–106.

[6] G. Jars, *Voyages metallurgiques*, 1774–81, vol. 2, p. 542; J. Wallis, *Northumberland*, vol. 1, 1769, p. 20.

[7] P. N. Wilson, 'The Nent Force level'. *Transactions of the Cumberland and Westmorland Antiquarian and Archaeological Society*, new series, vol. 63, 1963, pp. 253–80.

[8] 1864 report (vol. 2), p. 374.

[9] Ibid, p. 370.

[10] Ibid, p. 330.

[11] Jars, vol. 2, p. 543: 'There is very great negligence in this respect: I passed through many levels where the wood was rotten and broken, and was obviously dangerous.'

[12] Greenwich Hospital report, 1822, p. 8.

[13] 1864 report (vol. 2), pp. 374–5.

[14] Ibid, p. 361.

[15] Jars, vol. 2, p. 543.

[16] Sopwith, 1833, p. 129.

[17] Ibid, p. 70.

[18] 1864 report (vol. 2), p. 375.

[19] Ibid, p. 327.

[20] 1842 report (Leifchild), p. 682.

[21] Derwent mines report, 30 December 1872.

[22] 1864 report (appendix), p. 19.

[23] Many pages of Sopwith's diaries towards the end of his tenure of office as chief agent are taken up by a discussion of the potentialities of the rock drill.

[24] B/B 53, 30 June 1813.

[25] J. Granger, *Agriculture of Durham*, 1794, p. 20.

[26] There have been claims for the pre-eminence of both the Greenwich Hospital's Nent Force level and of the London Lead Company in introducing underground railways. But the Nent Force level was not started until 1776 and, as will be seen, the London Lead Company had no rails in its largest mine, Rampgill, in 1765. Rails must have been laid in Coalcleugh mine at about the same time as they were introduced underground in the Northumberland and Durham coalfield. (See R. L. Galloway, *Annals of coal mining*, 1898, p. 278.)

[27] B/B 98, March–June 1757.

[28] Jars, vol. 2, p. 542: 'When the ore and deads have been lifted through many shafts of different heights, either by windlass or gin, until they have reached the topmost level, they are brought through it to the day on roads similar to those in use at Newcastle for carrying coal, and in the same sort of carriages, but different in that instead of there being 4 feet between the pieces of wood on which they roll there are only two, and the carriages themselves are small and low: a single horse draws two of them when loaded. The level is continued a great way forward in the vein, as far as the boundary of the two undertakings: and as the London Company have not the same convenience for removing their rubbish, they pay the other company *for passage*, a fixed sum for every load sent along the level.'

[29] Apparently first in the Nent Force level shortly after 1805. Greenwich Hospital report, 1805, p. 251.

[30] Sopwith, 1833, pp. 131–2.

[31] G.H. (Wigan) MS.

[32] Blackett of Matfen papers.

[33] Vol. 2, pp. 363–8.

[34] Vol. 2, p. 490.

[35] Derwent mines report, 20 January 1872.

[36] Jars, vol. 2, p. 542: 'But to raise the water from a greater depth it is their intention to profit from that water which passes in great quantities through the upper level to work a machine of a new type the principle of which has been suggested by the fire engine. It is proposed to have a similar cylinder but instead of the column of air, the weight of the atmosphere, they intend to construct upon the piston a very great column of water.'

[37] W. White, *Northumberland and the border*, 1859, pp. 464–5.

[38] B/B 133, 22 October 1746.

[39] B/B 53, 24 October 1807.

[40] B/B 133, 13 October 1749.

[41] B/B 136, 9 February 1779.

[42] 1842 report (Mitchell), p. 727.

[43] W. Forster, *Section of the strata*, 1821, pp. 241–3.

[44] Sopwith, 1833, p. 130.

[45] 1842 report (Mitchell), pp. 726–7.

[46] In *Archaeologia Aeliana*, vol. 1, 1822, pp. 182–6. Gunpowder was in use from the late seventeenth century onwards.

[47] Sopwith, *Diary*, 30 October 1868.

[48] 1842 report (Mitchell), p. 762.

[49] 1864 report (vol. 2), pp. 330–1.

[50] B/B 53, 6 April 1807.

[51] 1842 report (Mitchell), p. 761.

[52] B/B 50, 23 January 1786.

[53] B/B 167, 7 January 1762 and 3 March 1762.

[54] B/B 176, 9 January 1749.

[55] B/B 48, 11 November 1764: 'They had 9s a week to pay for drawing water so they were obliged to quit that bargain or ruin themselves.'

[56] B/B 53, 6 April 1807.

[57] 1864 report (vol. 2), p. 331.

[58] B/B 131, 11 December 1860.

[59] B/B 167, 21 May 1760.

[60] Raistrick, 1938, pp. 55–6.

[61] 1864 report (vol. 2), p. 346; (vol. 1), p. 261.

[62] 1842 report (Mitchell), pp. 742–3.

[63] 1864 report (vol. 2), p. 332.

[64] B/B 54, Christmas 1818.

[65] 1864 report (vol. 2), p. 13.

[66] 1842 report (Mitchell), p. 761.

[67] 1864 report (vol. 2), p. 353.

[68] W. Robinson, *Lead miners and their diseases*, 1893, p. 4.

[69] B/B 54, Michaelmas 1819.

[70] B/B 176, 1 July 1749.

[71] Jars, vol. 2, p. 543: 'If the managers understood the theory of the circulation of air in mines it would be easy to expel this bad air by means of vents from one working to another.'

[72] Forster, 1821, pp. 322–3.

[73] 1864 report (vol. 2), pp. 350–1.

[74] Ibid, p. 338.

[75] Ibid, p. 369.

[76] W. Turner, 'Short tour through the lead mine districts, 1793.' *Transactions of the Newcastle Literary and Philosophical Society*, vol. 1, 1831, pp. 66–81.

[77] Sopwith, 1833, pp. 119–20.

[78] 1864 report (vol. 2), p. 377.

[79] Ibid, p. 367.

[80] B/B 136, Weardale, 25 August 1787.

[81] 1842 report (Mitchell), p. 728.

[82] B/B 53, 10 April 1809.

[83] B/B 176, 9 July 1747.

[84] 1842 report (Mitchell), p. 739.

[85] 1864 report (vol. 2), p. 360.

[86] Ibid, p. 371.

[87] Ibid, p. 13.

[88] Ibid, p. 379.

III

The bargain system

The relationship between employer and employee in the lead mines was governed by the bargain system. In theory, this contract between the two sides—capital and labour—remained the same throughout the eighteenth and nineteenth centuries. In practice, as will be seen, the interpretation of the bargain altered considerably. The theory and even the format of the bargains stood unchanged right up to the closing of most of the mines in the late 1870's. The interpretation of the basic contract, however, had changed considerably by 1842, and the changes continued at a more rapid rate after that date. The overall result was a general tightening of the mine agents' control over the workmen, so that the miners, although still nominally independent contractors, found themselves forced to work regular shifts at specified times, and many of their traditional holidays and celebrations quietly disappeared.

The nature of the sub-contract

The following quotations from various sources covering nearly a hundred years show how little the bargain changed in its basic constituents. Jars, writing in 1765, described the system:

Tous les ouvriers en général travaillent à prix-fait par troupe de deux, quatre, six, huit, etc. On leur donne tout par toise pour des endriots ou il n'y a pas du minérai, & pour ceux ou il y en a, le prix est fixé pour chaque mesure que l'on nomme *bing* (8 cwt.); sur ce prix ils sont tenus de se fournir les outils, la poudre, la lumière, de trier le minérai & de la livrer pret a etre fondu.[1]

In 1821 Forster wrote:

The miners take a certain piece of ground, commonly called a *length* in which they propose to raise ore, for a certain time, at so much a Bing, according to the richness of the Mine or working. A length of ground is commonly either twelve, fifteen, or twenty fathoms, and the price of procuring the Ore, depends much upon the hardness, the expence of drawing the Stone or Ore out of the Mine, and the probable quantity of Metal that can be raised . . ., the miners always paying for Candles, Gunpowder, the expences of drawing the Ore or Stone from the Mine, working, dressing, and preparing it, fit for the process of Smelting.'[2]

The 1842 report stated that:

The miners with few exceptions, speculate on the produce of the mines in which they work. Four miners . . . form a partnership together, and they make a bargain that they will work in a certain part of a certain mine for the next three months, for so many shillings for every bing of ore which they dig out: the expense of washing and cleaning the ore, and making it fit for the furnace, being charged to them. All the other men in the mine, in parties of 4, 5, or 6, or more make bargains in the same way. These bargains are entered into a book, and the miners sign them.[3]

Thomas Sopwith, on being examined by the 1864 commission, described the bargain system in much the same terms, adding only that a time clause had lately been introduced, compelling the miners to work regular hours, which had at first been much resented.[4]

For the mine owners the bargain system allowed piecework payment of their employees—payment by results. The more ore the men raised, or ground they dug away, the more they were paid. The system encouraged the miners to work hard, for their own profit was linked closely with that of their masters. In a group of resolutions voted by the court of the London Lead Company in 1810, No. 12 was 'That as few Wagemen as possible be employed, but that in every instance where it can be accomplished, the work to be done by the Peice, of what nature soever.'[5] The Blacketts and Beaumonts followed the same principle. Their mine accounts show that no men engaged in getting ore or driving shafts and levels were paid wages. The 'wagemen' were those in the subsidiary occupations, blacksmiths, woodmen, etc., and formed a small proportion of the whole labour force.

The bargain system also allowed the owners to spread the risks of lead mining. The miners had to speculate in the same way as their masters, with the same chances of ill luck. But the miners working in a rich vein were paid far less per bing of ore than those working in a poor one. The richer the mine, the less the owners had to pay per bing. The workmen, therefore, did not share the good fortune to the extent that they shared the bad. If, for example, the price of lead fell to a level which made it uneconomic to work poor veins, mining operations were ruthlessly curtailed and part of the labour force would be dismissed.

Another resolution agreed to by the court of the London Lead

Company in 1810 was 'That in letting of Bargains no person whatever shall have an hireling employed on any account in the Company's Works'.[5] In other words, every partnership had to work its own bargain, and not hire other men to do the work on a wage basis. This had apparently been a common practice in the Blackett/Beaumont mines until Sir Walter Blackett put a stop to it in 1764. The mine agents had been giving bargains to their 'hired servants', a practice that could obviously lead to corruption. Sir Walter wrote to one agent who wanted to do this:

After you had left me the other day, I considered the nature of the request you had made to me and which I agreed to upon your telling me it was an usual practice, to wit, to put a Man into the Grove to work for you; that is, you agree to give him a certain daily Sum and all the Oar he gets is to be your's. Now according to my notion this agreement cannot be for my advantage, a man cannot be supposed to work so diligently for another as he will for himself (witness all the Statute work which is done upon the Roads) wherefore I have sent to my Stewards, that no such persons, who do not work for their own profit entirely, shall be taken into the Groves.[6]

Instructions to the Blackett/Beaumont agents issued at the beginning of the nineteenth century enjoined them strictly that only such persons whose names had been 'inserted in the Bargain Book at the time of contracting, to be entitled to the benefit or advantage arising from such Bargains.'[7] In 1813, after the chief agent had observed 'in the Coalcleugh Accounts for the last year . . . the very large sum of £1147–2–3 earned by 2 different partnerships both under the name of Joseph Walton', he instigated an enquiry resulting in the Coalcleugh agent's dismissal. The circular letter to the other agents explaining this act ordered them 'to avoid in every instance partiality by observing an equal consideration towards all the workmen and to discontinue the improper practice, which has been excercised at Coalcleugh contrary to the positive injunctions laid down for your Regulation, of employing Hirelings'.[8] After this episode there is no later evidence of corruption or collaboration between agents and men in the Beaumont records.

The contracting miners liked the bargain system for the independence it gave them. They felt themselves to be free men working for their own profit at times of their own choosing. This freedom was to become more and more illusory as the nineteenth century progressed. The administration of the system is well illustrated by the Blackett/Beaumont records, where long series of bargain books cover every one

of their mining districts. Information about practices of other companies is far more scanty, except for the evidence gathered by the
1842 and 1864 commissions.

The bargains were conceived to have the same legal nature as
leases. In the eighteenth-century bargain books covering the Blacketts'
Weardale area, 'tacks'—or leases of ground to small groups of adventurers giving them full mining rights for a period of years—are
recorded alongside the normal bargain entries. Bargains were contracts
between the employers and the employed, with legal recognition. In
1805, after a threat of a strike in the London Lead Company's mines,
the magistrates 'compel the Miners to complete their Bargains.'[9]

The bargains were recorded in 'bargain books'. The format varied
in the Blackett/Beaumont mines from place to place and period to
period (see plate III). The fullest form of entry included a list of all the
partners, a definition of the area of work, the price to be given for
work performed, and the period of the agreement. Sometimes one of
the partners signed the record. A duplicate of the bargain contract was
given to the partnership concerned, and the Beaumonts introduced a
printed form for this purpose in the 1860's.

In the eighteenth century the bargains of the two largest concerns
ended, or were renewed, at three-monthly intervals. The Blackett/
Beaumont records show that most bargains were taken in January,
April, July and October. There was some unevenness of letting; most
bargains were let at the beginning of the quarter but some not until
the middle or end of it. Nearly all ended at the same time. The bargains
for longer periods were mainly of the 'tontale' variety (to be explained
later) or those at small, isolated mines with only one or two partnerships working. Occasionally there was a bargain allowing the partners
discretion when they were to finish: 'to June 1754, but if it prove a
hard Bargain they're to be loose at March 31st 1754.'[10] In the
nineteenth century, letting was more regular, and both the larger
concerns changed to a four-monthly bargain period. The court of the
London Lead Company resolved in 1810 'That no Bargain be let for
a longer Term than a Quarter of a Year.'[11] In 1816, however, the
bargain period was lengthened to four months, so that bargains were
taken only three times a year instead of four. The Beaumont concern
retained a three-monthly bargain period until 1859, when it too
changed to a four-monthly period.

'Letting' the bargains was an elaborate ritual which required complex organisation. In 1845 Sopwith described in his diary the procedure

involved at Allenheads (which had apparently changed little in the preceding century).

26 Sept. I went at an early hour to accompany the Inspectors of Mines on their quarterly examination of the several workings preparatory to arranging the prices for new contracts. This survey is called 'Riding the yards' . . . *4 Oct.* I let the Bargains . . . This occupied the greater part of the day and was followed by a dinner at the Inn at which all the Inspectors and chief clerks were usually present and which I was expected to provide . . . *6 Oct.* This is called 'Placing day' . . . Some workmen when offered a Bargain . . . refuse the offered price and if no agreement is come to, they remain 'out'— or unemployed. Occasionally there is not employment for the entire body of miners and those who are not required, are also 'out'. The object of 'Placing day' is to reconsider the contract prices and to endeavour to find occupation for all the workmen resident in the district.

In 1864 the system had changed little. 'Viewing agents' inspected the mines and recommended a price. The actual letting was done in the presence of the agents of Allenheads, Coalcleugh and Weardale, as well as of the chief agent. 'We give independent opinions, so that we cannot cheat anyone.'[12]

The London Lead Company's chief agent gave the 1864 commission a description of the procedure in the company's mines.

We send not fewer than five agents to each working, throughout the whole of the mines . . . Formerly one set of agents used to view all the mines, and it took them a fortnight to complete their work; . . . but for some years past, instead of having one set of agents going, I have had two sets to view, and we are able to accomplish all our bargain arrangements within a week. As a general rule, the men do not work at the time that the view is going on, so that we have a loss of time, a loss of time to the masters, and a loss of time to the men also . . . . These agents are not exclusively the local agents, but they are drawn from the different districts . . . mingling the home agents and the stranger agents together; and in that way all possibility of partiality, and all possibility of any improper dealing between the local men and the miners is prevented.

This system had been in operation 'for the last half century'.[13]

The development of the system

The Blackett/Beaumont bargain books provide a splendid record of the changes in the conditions imposed for a period of over 150 years. In the eighteenth century there were four basic kinds of bargain. The

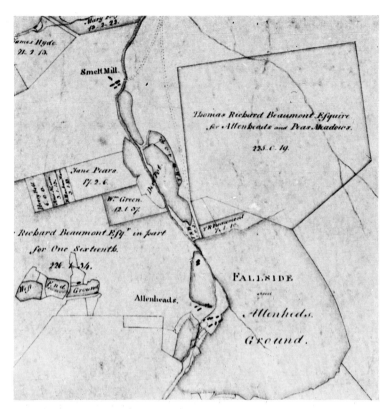

Plate V   Allenheads in 1793. *From the Allendale parish enclosure map*

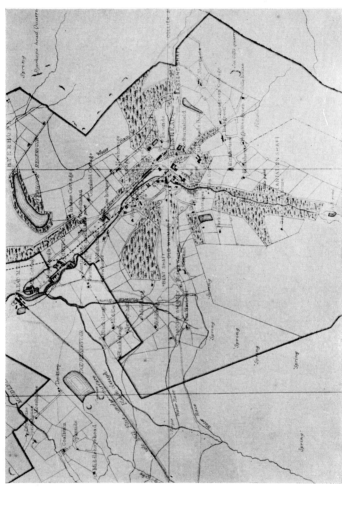

Plate VI   Allenheads in 1861. *From a privately printed survey of Allenheads*

most usual was 'bingtale': the mining partnership was paid according to the number of bings of *ore* it extracted from the mine and washed ready for smelting. Then there was 'fathomtale': the partnership was paid according to the total number of fathoms it had driven through the ground. 'Tontale' was payment based on the total number of tons of *lead* produced from the ore extracted by an individual partnership. Lastly there were 'lump bargains', where the partnership had been given a task or tasks to do for an agreed payment. In practice, of course, there were many variations and combinations of these basic forms.

In the very earliest surviving bargain book, for Weardale in the 1720's, the agreements are mainly of the first two types. By the middle of the eighteenth century, variations were common. Frequent changes of practice suggest that the agents in the early and middle eighteenth century were not given such precise instructions as to the conduct of bargains as their later successors. A change of agent led to a change of bargain practice, since each ordered matters as he chose. In the 1750's lump bargains were more common than fathom contracts. Unpleasant conditions were frequently imposed. 'Let . . . a Bargain to Cut Cross to a Sun Vein in the Low Hazel Sill . . . at £15 per Lump: but if it so happen that they do not Cutt a Vein they are to have nothing for Cutting Cross.'[14] Differential prices based on the number of bings raised, which had been common earlier, disappeared. As an example, a typical bargain of the 1720's altered the price if the men were too successful: 'to get ore at 20/- per bing, but in case they gett above fifteen Bings of ore before the 31st of Oct next they are to have but 18/- per Bing'.[15]

The reason for the existence of tontale as well as bingtale bargains for the extraction of ore is not obvious. In the early eighteenth century, and for all of the nineteenth after about 1810, tontale bargains were limited to the workers who were engaged in washing the 'wastes'—the heaps of deads and rejected material—for the scraps of ore remaining. In the case of these 'Waste washers', the owners preferred tontale bargains, as the ore won from this further washing was often inferior to that coming direct from the mines. In the late eighteenth century, and particularly in the period 1790–1806, tontale bargains were taken by the ordinary underground workers as well. At Allenheads in January 1793 more than half the total bargains were tontale. One possible reason why the mine owners favoured the ton of lead rather than the bing of ore as the basis for payment was that it prevented the men from cheating by retaining impurities in the ore to add to its

weight. In 1732 the chief agent had complained about the quality of
the ore received at Dukesfield smelt mill from Allenheads, 'and particularly what was sent from Allison Shaft'. The ore had been very wet,
probably 'occasioned by the Bargs. throwing Water upon it to make it
weigh'.[16] In 1802 some of the Weardale bargains gave the agent the
choice of paying the men by bing of ore or by ton of lead. 'To get
Ore . . . at 42/– per Bing till Dec. 31st 1802. N.B. if their Ore do not
please the Viewer they are £11 per Ton.'[17] Tontale bargains for the
ordinary miners disappeared as mechanisation was applied to washing
methods from the first decade of the nineteenth century onwards, and
by 1820 tontale bargains were again taken only by the waste washers.

The washing process was another feature in the bargain organisation
that was to change. According to the wording of the bargain the contracting partnership was responsible for washing the ore as well as
extracting it. For most of the eighteenth century each partnership did
wash its own ore, with the help of the sons and sometimes the wives
of its members. The equipment was very crude and primitive. Around
the turn of the century, the two largest companies, inspired by Cornish
practices, introduced new machinery which made washing by the
individual partnerships uneconomic and comparatively inefficient. By
the 1830's washing floors with machinery driven by water power had
been installed at all the large mines. These floors were managed by the
companies, and the miners spent all their time getting ore. A charge
for each bing of ore washed was deducted from the money earned by
each partnership.

The nature of the fathomtale contract changed too. In the eighteenth century—and in the case of the Beaumont mines, until well on
into the nineteenth century—fathom work implied merely the removal
of barren ground. Any ore found in the process belonged to the workmen, and they were given a price for it. The terms of the contract
often recognised this—a partnership received so much per fathom for
ground cut, and so much per bing for ore raised. Later it became
accepted in some concerns that ore extracted by fathom workers was
the property of the mine owners. In 1817 Robert Stagg, London Lead
Company chief agent, reported to the court of the company that he
was conducting 'an experiment to ascertain whether it may not be
more advantageous in some part of this Vein to cut the whole of it out
by the Fathom instead of working it by the Bing. By the Mode now
trying much Wood will be saved, the Wages will be more equalised,
and there will be considerable saving also in the Washing [this saving

in the washing was due to the fact that by the new system the 'bouse', or unwashed ore, extracted by each partnership no longer had to be kept carefully separate] but it remains to be proved by trial which will be most profitable.'[18] In 1842 both systems of fathom payment were in use. In the Beaumont mines the old system prevailed; in the London Lead Company's, the new. In 1864 the chief agent of the London Lead Company told the commissioner that payment by bing and by fathom were both still in use, but that 'three fourths' of company's miners were now paid by fathom.[19]

A few years before 1864 the Derwent mines had been taken over by a new company which had introduced Cornish customs, Cornish terminology and a few Cornish miners. The agent of these mines talked to the 1864 commissioners in Cornish terms—bingtale was called 'tribute work' and fathomtale 'tutwork'. Instead of the letting of bargains there were 'settings'. The procedure was completely different from that elsewhere on the mining field. 'Sometimes we let our bargains for a month, but generally speaking, we give them a certain stint, two or three fathoms, and they may have a second or probably a third bargain too in the month.'[20] This was normal Cornish tutwork practice. A surviving report book of the Derwent company from the 1870's shows that the Cornish system of bidding for 'sets' was in operation, the men competing with one another to take favoured sets at a percentage of the value of the total ore raised. The tributer who was prepared to take the lowest percentage was allowed to work the set.

In the small mine at Stonecroft, north of Hexham, bingtale had been abandoned altogether by 1864. Men were paid 'by the square fathom', i.e. six feet high and six feet forward.[21] In the Beaumont mines the old arrangements continued longer than in those belonging to the other mining companies. The system was streamlined— for example, forms were specially printed to serve as the miners' record of bargains—an eight-hour day was enforced, the period of bargains was extended to four months, and there was virtually no stoppage of work at the beginning and end of the bargain periods. But as late as 1875 the Weardale bargain books show an almost equal number of fathom and bing contracts. By the end of the 1870's, when lead prices had fallen to a disastrous level and the Beaumonts were considering abandoning mining altogether, the book-keeping degenerated sharply. Bargains were badly entered and, if renewed, the new dates were written in over the old, instead of the entry being rewritten.

The essence of the whole bargain system was the fixing of the length to be worked, and the price to be given. The principles governing these vital factors remained the same all the time the bargain system was in operation, but as the nineteenth century progressed so did the agents' control over the deployment of the workmen.

The operation of the system

Even in the early days, the miners were not allowed complete freedom as to where they were to work. At Fallowfield in the late seventeenth century, each partnership was allocated a shaft from which to dig ore. Directions to the Blackett miners in the eighteenth century were normally precise, e.g. 'to get Ore in the ground, East of the hole that Bainbridge's Horse fell into'.[22] If a partnership raised ore from a part of the mine or part of a vein without a contract to do so, its members were not paid for it.[23] The reason was, of course, that the price a partnership was paid was based on the precise conditions at a particular place. The factors that both the agent and the contracting miners had to consider included the hardness of the vein, its quality, whether blasting was necessary or not, the distance from the surface and the nearness of a horse level, and the ventilation and drainage conditions. If the vein were rich and access easy, a relatively low price was given; if the opposite, a much higher price per bing. The price had to be agreeable to both sides. Over and above these local factors, the agents had to consider the prevailing economic conditions—the price of lead, and of food—for even when the mines could be worked only at a near loss some work had to be kept going or the skilled labour force would migrate from the area.

The miners were 'free to accept or refuse the price the Agents . . . fix upon the different Workings'.[24] They could suggest to the agent where they would like to work, but more usually it was done the other way about. However, at the bargains, the partnership at a given place expected to work it again if its members chose, and a price could be agreed on. One Allenheads partnership wrote a letter of complaint to Sir Walter Blackett when the local agent 'as soon as their Bargain was out, . . . took the best of the ground from them where she was three-quarters of a yard clean Ore in width and let at 17/– a Bing, and put them into the worst ground'.[25] If the original partnership refused to renew a bargain because its members thought the new price too low, another partnership was free to move in if it was prepared to accept the price. A low and uneconomic price, wrote Sopwith in 1852, would sometimes be taken 'with a view to obtain a footing and to displace others'.[26] The mine owner always gained the advantage in this sort of manœuvring as, save in the rare times of great demand for lead, there was almost invariably a surplus of labour over jobs available.

Detailed reports covering the years 1816 to 1819, written by the London Lead Company's newly appointed chief agent, Robert Stagg, illustrate how control over the mine workings was maintained, and a balance held between extracting ore and doing the necessary dead work for future expansion. In 1816 Stagg wrote that the prices in one mine were far too high, 'the chief workings having been pushed forwards upwards of 20 Fathoms before the drawing Level from an improper anxiety on the part of Mr Dodd [the previous agent] to raise a great quantity of Ore'. The men were temporarily removed from these forward workings to the 'old pickings' to allow the necessary dead work of driving the horse level to progress. Stagg calculated that the average price per bing would then drop 7s. Men working in the old pickings were given very high prices as these were 'parts of the Mines which had been left as being too poor to work & are all therefore entirely speculations on the part of the men'. But 'it is a notorious fact that a great proportion of the good Mines now working were originally discovered by speculations of this description made by the Men, rather than that they would be entirely out of work'.[27]

A report of 1819 shows how the miners were deployed in their attack on a vein. Each partnership was allotted a square, 'which they cut round in the first instance, whereby the Agents are enabled to judge with comparative certainty of the prices that ought to be given'. These large squares would then be divided into 'two or more compartments' as was thought fit.[28] This regular method of approach to a vein was apparently introduced into the northern Pennine ore field by Stagg, and it replaced, in the larger mines, the random following of a vein in all its 'twitches' and 'throws'. By approaching in a pattern of squares, no ore was missed, and minor veins, or 'strings', were more certainly discovered.

It does not appear that the partnerships of miners were specialists in either bing or fathom work. The Blackett/Beaumont records in both the eighteenth and the nineteenth centuries show that a partnership would often change from bing to fathom work, or vice versa, at successive bargains. This fluidity of labour was essential to deal with the constantly changing conditions in a lead mine. A detailed account of the work of a single partnership at Fallowfield in 1770 shows that in a six-monthly period its members had raised twelve bings of ore, driven twenty-five fathoms, spent twenty-one days repairing levels and eight days in Fallowfield smelting mill.[29] This is an unusually varied example, due to Fallowfield being small and isolated. It is most unlikely that miners elsewhere did all these different jobs in a single quarter, but over the field as a whole there was none of the specialisation in either tributing or tutwork that was a feature of Cornish mining.

The number of men in a partnership varied from two (which was unusually low) to twelve; at all periods, as witnessed by the Blackett/ Beaumont records, most partnerships consisted of four to eight men. Every partnership had to declare its members at the time of the bargain, and was forbidden to hire any 'wagemen'. The only exception to this rule was for boys, usually the children of members of the partnership, who were taken into the mine by their fathers during the winter months. Very large partnerships of ten or twelve members were employed only at driving a level or a shaft where speed was essential, or in places where there was much water and constant pumping was required. In a report of 1816, Stagg wrote that one reason why it was economically better that a horse level should be close to the ore workings was that the large sizes of the partnerships could be reduced. 'Small partnerships of four men instead of the large partnerships of eight or ten men' could work at lower prices, as most of their time would be spent in the actual getting of the ore. The large partnerships had used much of their labour force in transporting ore and deads over a long distance by hand.[30]

The smaller partnership groups consisted of 'shoulder fellows, between whom a lasting intimacy springs up.'[31] The Blackett/Beaumont bargain books show that the smaller partnerships were fairly constant in their composition; the larger ones were *ad hoc* groupings made up as occasion demanded. Many of the partnerships were between members of the same family, often of different generations, 'and there is the skill of the one and the physical energy of the other, the one compensates for the other'.[32]

The employment of boys inside the mines was decreased by the introduction of long levels. Before this 'system of mining' came into use just after the mid-eighteenth century, boys were extensively used to convey ore and deads to the foot of the whimsey shaft. 'The work was brought to the drawing place along the Drifts by a number of

Boys, who had their Stages fixed.'[33] Horse-drawn waggons on rails greatly reduced the need for this sort of labour. By the nineteenth century the number of jobs that could be done by boys who had not yet come to their full strength was limited. They could help in getting the ore and deads from the mine—leading horses, loading waggons and kibbles, breaking up deads, and generally sweeping up. They were also employed in turning the wheels to drive fans in otherwise un-ventilated workings. The actual work of ore getting and cutting new levels was beyond their strength and skill, although they could assist the partnerships in the more menial tasks. In 1842 there was none of the wholesale exploitation of child labour found in coal mining. The reason, however, was economic, not humanitarian. Child labour could be employed more efficiently on the washing floors where the ore was dressed.

In 1842 there were only forty-one boys under 18 employed in the mines in East and West Allendale, eight of them under 14, the youngest being aged 10.[34] At that date 10 appears to have been the earliest age at which boys were allowed into the mines, although there were younger boys working on the washing floors. One of the medical witnesses to the 1864 commission enquired of miners when they had first begun working underground. In Weardale and in the Allendales many of the older men had begun work at the age of 7, at Nenthead 9 was the earliest.[35] For virtually all of them, however, such work was only temporary, when weather conditions had stopped work on the washing floors. By 1864 boys were not allowed into the mines until they had reached a certain age, which varied with the different companies. The London Lead Company allowed no one to work under-ground regularly until he had reached 18, although boys of 14 years and upwards were allowed in for three months during the winter.[36] Thirteen was the youngest age in the Beaumont mines, although Sopwith admitted that 'It is found practically difficult to enforce a rule rigidly. When family matters or other matters of that kind come into consideration, we sometimes relax the rule.'[37]

Most boys came into the mine only in the winter months when the washing floors closed down. In 1842 only the 16 and 17 year olds normally worked underground all the year round. They performed several tasks, and each type of job had its own form of payment. Those engaged in working ventilation machinery were employed on a wage basis by the mine owner. The boys removing ore and deads were employed by the contractor who owned the horses working in that level. The remainder were employed directly by the partnerships. In the eighteenth century these latter were mentioned in the Blackett/ Beaumont bargain books as '— & lad' and counted in the total number of partners, although they were not apparently paid as such. By the middle of the nineteenth century they were employed without being numbered as partners at the bargains. In 1842

when boys are first taken into a mine with men they are hired, and either paid wages by the shift, or so much in the shilling of the men's share of the produce of the contract when the ore is washed and weighed . . . . The wages of the boys employed in partnerships are paid by the men employing them: the wages are agreed for by their parents generally, but are frequently determined by age, and the agents of the mine sometimes fix the boys' wages.[38]

The boys worked the same hours as the men—hardly ever more than eight hours a day.

The boys received a few pence a day; for example, a 14 year old would normally receive 9d a day in 1842. Eighteen was the earliest age at which a boy was usually admitted as a partner; 'if his father or brother be already working in the mine his advance to be a partner is much facilitated'.[39] Some were not admitted to full partnership until they were 20 or 21. Working with the same partnership, and eventually gaining acceptance by its members, was the way for a young man to become a full partner.[40] In a contract for drawing work at Allenheads in 1846, it was recorded that 'the undertaker to be assisted in hiring Lads to drive waggons etc., and to have two of the Lads (most deserving) taken annually from the driving and employ'd otherwise in the Mines'.[41]

The number of non-bargain adult workers in the mines were limited as much as possible. Basically, the day-rate men were of two kinds. First, there were the labourers doing odd jobs underground and on the surface. In the Blackett/Beaumont mines the accounts show that these workers were drawn from the men who were out of bargains for one reason or another, such as failure to come to an agreement over price, or their bargain proving so bad as to be unworkable. There was no regular class of unskilled labourers. The other type of day-rate man was the tradesman doing skilled work subsidiary to the main task of ore getting. The Blackett/Beaumont accounts show that the proportion of men employed by day-rate did not increase in the nineteenth century. If anything, the opposite occurred. From the 1830's onwards, complex bargains appear in the accounts by which the tradesmen contracted to do their work in the same way as the ore getters, though usually for longer periods. There were such bargains for blacksmiths, engineers and woodmen who had previously been paid on a wage basis. An example of such a complex bargain is one by which pumping was undertaken at a given place for one year:

William and John Curry agree to pump the water at the east end in Wentworth Vein, they finding Horse, driver and Candles—And to have the pump Horses shod, the Engine kept in repair at the expense of Mr. Beaumont for one year . . . if it is necessary to work the pump for that period, if not, to be paid in proportion for the time it is wrought, for the sum of Sixty Five Pounds for one year. If the quantity of water should increase to be more than one Horse can possibly keep, then the time of an additional Horse to be allowed, at the rate of 4/– per Day, including the driver, calculating six hours pumping for one day's work, but no additional Horse to be allowed, so long as it is found that one Horse can do the work, averaging seven hours per day through the week. N.B. The above W. J. Curry to take Mr. Beaumont's Horse which is now used for pumping at the value of Ten pounds.[42]

Hours of work

One of the most important changes in the organisation of lead mining was the transition from irregular hours and days of work in the eighteenth century to an enforced eight-hour day, five-day week by the middle of the nineteenth. This development is best chronicled in the Blackett/Beaumont records. The normal eighteenth-century bargains for ore getting make no mention of any fixed hours to be worked by the partnerships. It was apparently left to the men themselves to decide whether they worked long or short hours, the whole quarter, or merely a fraction of it. The accepted number of hours per day was probably six. A wageman's bargain of 1749 named six hours as being the normal shift.[43] In the London Lead Company, conditions were the same. An 1816 report mentioned 'the usual Miners week of five shifts of 6 hours each'.[44] Mackenzie, describing Northumberland lead miners in 1826, said that they 'seldom work more than six hours a day, which is called a shift'.[45] But if a partnership 'struck it rich', its members might work very much longer hours in order to get out as much ore as possible before the end of the quarter brought about a revision of the bargain price.

Nor was the number of days worked in a week strictly enforced. A Cumberland historian of 1800 said of Alston Moor miners that 'They work hard about four days in the week, and drink and play the other three'.[46] On the other hand, the inhabitants of Wolfcleugh in Weardale were especially commended by the Beaumont agent in 1811, as they 'go regularly to their work every day, Saturday not excepted, and in general work the whole of the day'.[47] Even later in the century, however, Saturday working was most unusual.

In exceptional cases the bargain included a time clause. In the eighteenth century this was invariably either for dead work, when the completion of a new level or shaft was urgently needed, or for pumping, which often had to be maintained for twenty-four hours a day, seven days a week. In these cases, large partnerships were employed, the men working in shifts and relieving one another. Sometimes a penalty clause was inserted in the bargain. A partnership of twelve men at Coalcleugh in 1821 had to relieve 'each other at the forehead, from Monday morning till 6 o'clock on Saturday evenings, if a single instance can be proved of the forehead standing in the above mentioned time an hour, to forfeit 10/– per fathom'.[48] Such shift work was unusual. The members of a partnership normally all worked together in a single shift, which had the advantage that the sixteen or eighteen hours between shifts allowed the air to clear of dust and powder fumes.

As for the actual time of day when the shifts were worked, about 7 a.m. was probably the usual time for entering the mines. At the beginning of the eighteenth century the Allenheads miners were invited to attend prayers 'every Morning at six o'clock, before they begin Work'.[49] In 1842 most miners still began work at 7 or 8 a.m.

In the decade before 1850 the old laxity of timekeeping disappeared in the Beaumont mines. The reason for this change was chiefly managerial desire for higher productivity but it was linked with the gradual improvement in ventilation practices, which allowed the veins to be worked for longer and more regular periods. In 1842 a new formula was added to bargains at Allenheads and other Beaumont mines: the partnerships were 'to work five days per week per man'. In 1843 this became 'to work five eight-hour shifts per week per man', and the latter remained a normal addition to bargains in the Beaumont mines until their closure. As late as 1841 the Allenheads agent had stated that the miners 'generally work . . . 6 hours in each day',[50] so there was a definite change of policy at this time. When Thomas Sopwith took over the agency in 1845 he increased the monthly subsistence money paid to miners. He told them on this occasion that 'the advance of forty shillings has reference to actual work performed during five days of eight hours each.[51] A clock tower was erected at

Allenheads so that the miners had no excuse for not keeping regular hours.

A strike there in 1849 was largely due to the enforcement of regular hours (see chapter VI). In a letter to the *Newcastle Guardian*, defending his actions against the accusations of the strikers, Sopwith wrote that eight hours per day had been 'for many years before I came' the accepted period of work, and that bargains were not just piecework, but also contracts to work a certain period of time.[52] The bargain books show, however, that although the time contract was written into bargains before Sopwith's arrival, the practice had only started two years before. Surprisingly, the 1849 strikers did not emphasise the novelty of a time clause in their contracts. They protested most against the 'spy system' Sopwith had introduced, whereby the agents observed when the miners entered and left the mine. The men were expected to be actually working for eight hours, the time necessary for reaching and leaving the face being additional to this. However, the strike was crushed; regular hours of working were fully established in the Beaumont mines from then onwards.

In the London Lead Company mines, five eight-hour shifts a week were already normal practice in 1842.[53] In 1864, and probably in 1842 also, the Teesdale miners, who while at work lived mainly in lodging shops at the isolated mines well above the habitable parts of the dale, were allowed to work four ten-hour shifts, and thus have a weekend of three days at home.[54]

In the smaller independent mines, too, discipline was becoming stricter by 1842. A miner at one of the smaller Alston Moor workings said in 1842, 'We have just time enough to get our dinner and then go to work again. If we were to sit too long at dinner, and master should know it, he might discharge us, and work is so scarce we shall not know where to go.'[55]

The actual hours of work both in 1842 and in 1864 were generally from 6 or 7 a.m. to 2 or 3 p.m. As in earlier times, miners would work longer hours or in alternate shifts if they struck a lucky bargain or were doing urgent dead work. Saturday work was uncommon and work on Sunday unknown.

Holidays and ceremonies at the letting of the bargains, the pays, etc., formerly numerous and sometimes extending over several days, disappeared or were limited to a single day. One custom that did persist was to allow the men 'two or three weeks from the mines' to get in the hay harvest in July or August.[56] Thomas Sopwith much disapproved of this 'period of laxity' which 'so much interferes with work in the Mines and with attendance at school'[57] but even he did not attempt to change the custom.

Areas of friction between men and management

The bargain system was devised, or rather developed, in a pre-industrial age. Its survival was obviously precarious in the rapidly changing social conditions of the eighteenth and nineteenth centuries. However, its basic framework persisted until the collapse of the industry. Points of friction did arise and caused many changes in detail—generally in favour of the masters. The enforcement of regular hours of work was one such friction point; another, the transformation of irregular and precarious earnings into what amounted to a regular system of wages, will be dealt with in the next chapter. Two more concerned the removal of ore and deads from the mines, and the provision and owner- ship of tools, candles and gunpowder.

All the ore and a considerable proportion of the deads dug in a mine had to be brought to the surface, and this was the responsibility of the partnership which had mined them. In the early days there was little administrative difficulty, if much hard labour. The material was dragged or hauled to the surface by each partnership working in- dependently. The introduction of the horse level and the great enlargement of mines from the mid-eighteenth century onwards created new problems. The horse levels, and to a lesser extent the whimseys, or horse-driven windlasses which hauled material up the shafts, could not be manned or 'horsed' by each partnership separately. Organisation was necessary to regulate the time when each partner- ship's ore and deads could be removed, and to provide the necessary horses. The waggons were provided free by the mine owners; each partnership in turn had to hire the labour of a horse. In 1821 'The expense of drawing the Ore or Stone, out of the Mine, when Horses are employed, is pretty considerable, and depends much upon the length of the Level or Adit, and depth of the Mine. In Alston Moor, it is usually drawn at so much per Shift, and at some mines a shift con- tains eight Waggons, at others only six . . . The expense of drawing a Shift . . . varies from 3/6, to 8/–, including the filling, driving, and emptying the Waggons, there being no allowance made for the differ- ence of weight between the Ore and the Stone.'[58] These charges were deducted from the miner's pay.

The ownership of the horses and control of the drawing work was organised in one of three ways. In the early days of horse levels, it was the prerogative of the agents. The second method was to let the drawing work to a contractor, in much the same way that bargains were let. The third method was for the mining company itself to own the horses. In all three, it was in the company's interests to see that its miners were not overcharged. The first method was dropped because of its liability to abuse and the second or third employed instead. If the contract system were used, the prices the contractor was to charge were laid down by the company.

The first mention of the problem in the London Lead Company's records occurs in 1785, when the court of the company ordered 'That no Agent ... be concerned in Whimseys or Letting Horses to draw Waggons, or Wood, or anything of that kind for the Company'.[59] Before this date the agents had been allowed to own and manage horses in the mines under their control, and the company's chief agent, Thomas Dodd, put in a pay claim in 1793 on the grounds that he had been promised 'an equivalent for Horses that former Agents used to employ'.[60] The contract system was still in use in 1810, when the court resolved (apparently in accord with contemporary practice) 'that for engaging of Whimsey or Level Horses used by the Company, Notice to be given to the parties letting the same for receiving proposals of the yearly Contract; the proposals to be given under seal and forwarded to the Superintendent'.[61] In 1816 the contract system was abandoned, and the company took over the ownership and organisation of all horses used in drawing work.[62] This system lasted until the closing of the mines.

The more extensive extant records of the Blackett/Beaumont mines show more of the reasons that lay behind these changes. By 1770 William Westgarth, the Coalcleugh agent who invented the hydraulic engine and introduced horse-drawn waggons into lead mines, was drawing a higher income per quarter from his horses than from his salary. A jotting at the back of a Weardale bargain book of 1788 indicates how the drawing was arranged. The horses worked a twenty-four hour day—the work of one partnership at '12 night', of another at '4 morning', and so on through to midnight again.[63] The 1796 petition from the Weardale workmen to Colonel Beaumont (see Appendix 3) shows that the miners disliked this system. They told the Colonel that 'were we to inumerate all the advantages resulting from the workes to the agent and his friends you would realy be astonished;

suffice it to say that the Money he receives for one horse employed in drawing the Ore and Strata amounts to the enormous sum of 120£ per annum and the number of horse's imployed in the mines at one third less price would aford comfortable livings to Several poor famleys'. The Newcastle chief agent commented to Colonel Beaumont on this petition that 'As to the charge of the Agent having the advantage of employing some horses, I do not see the evil arising from it, in case the charge be fair and reasonable, it has always been allowed to the Agents who can do it at a lower rate than you'.[64] The value of the 'advantage' to the agents was considerable. In 1799 for the quarter June–September the Coalcleugh agent's salary was £15. His income from his horses (before deduction of expenses) was £177 18s 6d.[65] In 1807, however, the agents complained that their drawing privilege was barely economic: 'They think the present prices being too little, not being any advance for upwards of 30 Years and both the Horses and their feed being considerably advanced since that time.'[66] Upon investigation, the chief agent found that the charges per shift were less than those of the London Lead Company, and he agreed that the charges in the Beaumont mines should be increased accordingly.

The prices the agents charged for drawing work were, therefore, strictly controlled. They could charge only a standard price which was authorised by the mine owner, and the complaint in the 1796 petition was more that the agents were overpaid than that the workers were overcharged. But the system was obviously open to abuse, and some time between 1807 and 1820 (possibly after the dismissal of the Coalcleugh agent for corruption and malpractice in 1813) ownership of the horses was taken over directly by the mine owners, the Beaumonts. In the bargain books, contracts were recorded with men undertaking the leading of the horses and the loading of the waggons. But for some reason direct ownership was abandoned again in 1833 at the time of a thorough reorganisation of many aspects of the Beaumont mines. Contractors for each mine were to buy all the horses at an agreed valuation and take over the drawing work. These contracts were renewed yearly and recorded in the bargain books. Among the points laid down in the very full Coalcleugh agreement were the price per shift the workmen were to be charged (it is noteworthy that this price was less than that charged in 1807), the conditions of employment of the wagemen leading and loading the waggons, and directions about keeping the produce of each partnership separate.[67] Similar agreements were entered into at the other mines, at Allenheads with the proviso that the workmen should have the 'privilege' of running their work to the shaft 'by their own labour without charge, if they so choose'.[68] These agreements continued without fundamental change, until the end.

The supply of tools, candles and gunpowder also caused difficulties. The lead mines were so isolated that the workmen had no choice but to get their equipment through the mining company; there was the dual risk of a form of truck developing and of the goods being inferior in quality.

The supply of tools caused little friction. The workmen bought their own—picks, shovels, hammers, prickers, etc.—from the company on starting work, or when they needed renewal. A regular item in the Blackett accounts throughout the eighteenth century was 'For so much pd. by the workmen for new geer and sharping'. In 1864 the system was still the same. The miners bought tools from the companies as they were needed, and the companies were always prepared to repurchase them at the same price, allowing for any loss of weight caused in use and sharpening. Some companies charged their workmen for sharpening, others did it free.[69] During the Allenheads strike of 1849 Sopwith was approached by some of the strikers who 'desired to have their working implements out of the Mine in order that they might thus be in a condition to go and seek employment in other Mines. These tools partly belong to the Owner of the Mines, and partly to the Miners (who pay for them by instalments) but overlooking any claim on the part of the Owner I at once agreed to consider the working tools as the property of the Miners and I gave them leave accordingly for them to have them'.[70]

The supply of gunpowder and candles did cause some trouble, though. Vast quantities of both were used—in 1864 the London Lead Company's miners used 113,000 lbs of candles and 93,000 lbs of gunpowder[71]—and their quality affected the efficiency, and even the lives, of the miners. Sopwith noted in his diary that 'sometime before I came the West Allen Miners seized a quantity of powder which they disliked and put it in the river'.[72] In 1809 the powder supplied to the London Lead Company's miners was so poor that 'the Workmen were obliged to dry it before the Fire, before they could use it, and it is as well that no misfortune happened by it'.[73] Probably a supplier had been increasing his profits by supplying cheaper goods. If he were the mine agent or someone connected with him it was easy for this to happen. The Blackett/Beaumont records show the difficulty of firmly suppressing this form of truck.

During Sir Walter Blackett's ownership 'Frauds and Irregularities arose from the Mine Agents furnishing Articles for the Mines, and it was found necessary to put a stop to that evil, since which it has been a standing rule in the Concern, as well as that of the Lead Company, that no Lead Mine Agent (whose duty it is to receive and examine the quantity and quality of each Article) shall furnish Mines with any one Article'.[74] This statement was contained in a letter of 1787 from the chief agent to the new owner, Sir Thomas Blackett, who had just allowed the son of the Allenheads agent to supply the miners with candles. The chief agent did not suggest that the decision should be rescinded but that another supplier should also be approached, as he hoped that there would then be 'an emulation between them which shall furnish Candles of the best quality'. Apparently no such happy result followed, for the Weardale miners complained in their 1796 petition that the candles they were supplied with were not only more expensive than 'those of which other miners use' but that they were of an inferior quality and 'much brok'. The gunpowder arrived lacking 'the wrappers in which the Barrels are contain'd for the safety of the Powder and Convenience of carriage which wrappers are delivered in barter at 8d per stone to sellers of earthen ware'. The Beaumonts did not finally get rid of this wretched abuse until after Sopwith's appointment in 1845. He discovered that the candles were then supplied 'by parties connected with the mines' so he transferred the contract to 'respectable firms at Hexham, Bishop Auckland and Richmond'.[75] The court of the London Lead Company had firmly laid it down in 1810 'That no Gunpowder (or candles) be used by the Company's Workmen but what is furnished by order of the Court & to be charged to them at prime Cost, with only the addition of Freight and Carriage'.[76]

None of the agents questioned by the 1864 commission had any connection with the supply of goods for the mines. A table in the appendix to the report[77] showed that it was almost universal for miners to be charged only cost price for candles and powder.

On the surface, then, the bargain system changed little. In reality it changed so greatly as to convert the theoretically independent contracting partnerships into *de facto* wage-earners. The impetus for the change came from the mine owners, seeking higher productivity and greater managerial efficiency. The ratio of agents to workmen increased considerably in the nineteenth century. Greater administrative regulation became possible. The miners accepted the changes, if

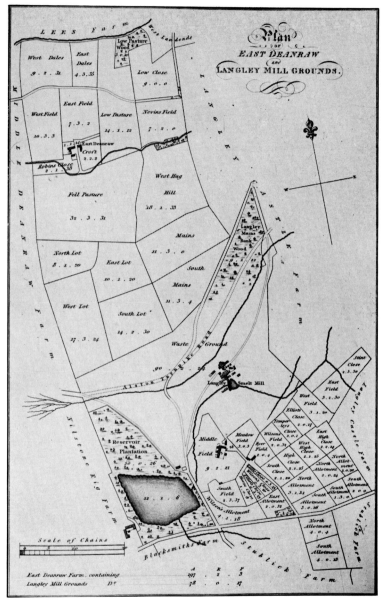

Plate VII   Langley mill grounds, *c.* 1805. *From a privately printed collection of engraved plans of the Greenwich Hospital estates*

on occasions reluctantly, in exchange for higher, or at least more regular, remuneration. The undoubted virtues of the bargain system —in encouraging initiative and hard work while preserving some feeling of freedom—allowed it to survive in spite of the pressures to which it was subjected. No less a person than Charles Darwin, commenting on the degradation of Chilean miners who spent their wages 'like sailors with prize money', thought that in the English metalliferous mines 'where the system of selling part of the vein is followed, the miners, from being obliged to act and think for themselves, are a singularly intelligent and well-conducted set of men.'

NOTES

[1] Jars, vol. 2, p. 543: 'In general all the workmen work at prices fixed according to the work they are doing in partnerships of two, four, six, eight, etc. So much a fathom is given them in places where there is no ore, and where there is, the price is fixed for every measure, which is called a bing. They are obliged to find their own tools, powder, and light; to dress the ore and deliver it fit for smelting.'

[2] Forster, second edition, 1821, pp. 332–3.

[3] 1842 report (Mitchell), p. 744.

[4] 1864 report (vol. 1), p. 262.

[5] L.L.C. 17, 6 December 1810.

[6] B/B 20, July 1764.

[7] 'Arrangements recommended to Colonel and Mrs Beaumont for the future agency and management of their lead mines,' Blackett of Wylam papers, appendix 2.

[8] B/B 51, 19 June 1813.

[9] L.L.C. 16, 16 September 1805.

[10] B/B 133, 21 January 1754.

[11] L.L.C. 17, 6 December 1810.

[12] 1864 report (vol. 2), p. 332.

[13] Ibid, pp. 375–6.

[14] B/B 133, 1 August 1754.

[15] B/B 132, 13 October 1724.

[16] B/B B.M. letter book, 16 May 1732.

[17] B/B 138, 1 November 1802.

[18] L.L.C. 16a, Midsummer 1817.

[19] 1864 report (vol. 2), p. 376.

[20] Ibid, p. 347.

[21] Ibid, p. 355.

[22] B/B 167, 21 October 1783.

E

[23] B/B 46, 16 January 1760.
[24] B/B 51, 26 December 1816.
[25] B/B 48, 11 November 1764.
[26] Sopwith, *Diary*, 16 April 1852.
[27] L.L.C. 16a, Midsummer 1816.
[28] L.L.C. 16a, Lady Day 1819.
[29] Fallowfield journal, 1769.
[30] L.L.C. 16a, Midsummer 1816.
[31] J. R. Featherston, *Weardale men and manners*, 1840, pp. 64–5.
[32] 1864 report (vol. 2), p. 379.
[33] G.H. (Wigan) MS.
[34] 1842 report (Mitchell), p. 746.
[35] 1864 report (appendix), pp. 20–1, 82–91.
[36] Ibid (vol. 2), p. 376.
[37] Ibid (vol. 1), p. 258.
[38] 1842 report (Leifchild), p. 681.
[39] Ibid, p. 745.
[40] Ibid, p. 760.
[41] B/B 64, 1 November 1846.
[42] B/B 62, 22 January 1833.
[43] B/B 176, 5 June 1749.
[44] L.L.C. 16a, Lady Day 1816.
[45] Mackenzie, second edition, 1826, vol. 1, p. 208.
[46] J. Housman, *Cumberland*, 1800, p. 70.
[47] B/B 53, 9 April 1811.
[48] B/B 81, 28 June 1821.
[49] G. Ritschel, *Account of certain charities in Tindale ward*, 1713, pp. 16–17.
[50] 1842 report (Leifchild), p. 682.
[51] T. Sopwith, *Observations addressed to the miners*, 1846, p. 15.
[52] *Newcastle Guardian*, 31 March 1849.
[53] 1842 report (Mitchell), p. 756.
[54] 1864 report (appendix), p. 18.
[55] 1842 report (Mitchell), p. 769.
[56] *Rating of mines:* select committee report, 1857, p. 7.
[57] Sopwith, 7 August 1856.
[58] Forster, 1821, pp. 33–4.
[59] L.L.C. 13a, 6 October 1785.
[60] L.L.C. 17, 7 November 1793.
[61] L.L.C. 17, 6 December 1810.
[62] Raistrick, 1938, pp. 91–2.
[63] B/B 167.
[64] B/B 50, 17 January 1797.

[65] B/B 98, 1799.
[66] B/B 53, 24 October 1807.
[67] B/B 81, 2 April 1833.
[68] B/B 62, 23 November 1833.
[69] 1864 report (vol. 2), pp. 331–2, 359.
[70] Sopwith, *Diary*, 22 March 1849.
[71] 1864 report (appendix), p. 442.
[72] Sopwith, *Diary*, 16 April 1852.
[73] L.L.C. 16a, 20 January 1810.
[74] B/B 50, 21 October 1787.
[75] Sopwith, *Diary*, 1 July 1865.
[76] L.L.C. 17, 6 December 1810.
[77] 1864 report (appendix), pp. 441–3.

IV

Payment and earnings

In spite of the span of the Blackett/Beaumont records, which include series of account books covering the greater part of both the eighteenth and the nineteenth centuries, it is almost impossible to give a meaningful account of the lead miners' average earnings. They worked in partnerships, and were paid on a piecework basis. Because of this, and because of the uncertainties of mining, their earnings varied enormously. The peculiar method of accounting and payment can be described in detail, but the amount of money that individual lead miners made in a year varied so greatly, both from year to year and from individual to individual, that any attempt to average it out is misleading. This chapter attempts to describe the administration of the system, and to give such facts as can be fairly stated concerning wage levels. Lastly, there is an account of the benefit societies that existed in the area to provide money for sickness and retirement.

We have seen that the bargain system was basically piecework. The miners were paid according to the number of bings of ore or tons of lead they raised, or the number of fathoms driven. The amount to be paid for working at a given job in a given place was settled in advance after an evaluation by both sides of the various factors involved. No human skill, however, could judge with certainty what would occur as working progressed; good or ill fortune vitally affected the eventual earnings of the miners. After every three or four months, conditions were reviewed and prices modified as necessary. The miners were paid once or twice a year. Periodically, between 'pays', each miner received a subsistence allowance which, together with money for candles, washing charges, etc., was deducted from his 'pay'.

In 1842 Dr Mitchell described the accounting system employed by the two major companies to calculate the exact sums due to their employees.

An account is opened against each miner, and he receives say 40s, which is called lent-money, on the first Friday in every month, which is entered against him. Also if any tools be supplied to him, or gunpowder, an entry is made against him. So also his quota of the expense of washing the ore. Then when the ore has been washed which the partnership have dug during the three months, the part of the money to which the miner is entitled is entered on the other side of the account to his credit. If the same partnership go on, then the proceedings of the second period of three months are the same as the first three months; and so it is with the third period of three months, and with the fourth period of three months. If the miner shall have entered into partnership with any other persons it makes no difference in the manner of keeping his account; after three months' ore has been washed his proportion is put to his credit. Suppose that the miner's year shall have terminated at Michaelmas, it will be some time before the ore shall be all washed, but when that shall have been accomplished the masters are now in a condition to make up his year's account. If the money which he has earned shall exceed the lent money of £2 every month and the other moneys chargeable against him, then there is a balance coming to him, and that is paid over.[1]

If there were a deficit at the end of the year it was carried forward to be deducted from the following year's pay.

Each miner, then, was paid a monthly or bi-monthly subsistence allowance, and received the remaining money due to him at periodic intervals. The chief source of evidence concerning the administration of these two forms of payment is the Blackett/Beaumont records, but those of the London Lead Company also give some related data; the 1842 and 1864 reports give some information about the nineteenth-century practices of the smaller companies.

The annual pays

The mining year for the Blackett/Beaumont concern ran from the beginning of October to the end of September.[2] This was the period for which the miners received an annual pay. In the eighteenth century the pay was often much delayed, and held at irregular times during the year. Sometimes a period of two years after the end of the mining year might elapse before the miners received their money. A letter of 1756 records that the pays would be held 'as usual, sometime in September'.[3] By 1793 'the usual time of making the Lead Pays has been the last week in April'.[4] In 1812 the miners of Coalcleugh complained when the pays were delayed until May instead of 'in or about the first week in March as usual'.[5] By 1842 the Beaumont miners were paid 'generally in the month of January',[6] and Sopwith's diary shows that from 1846 to 1860 pays were regularly made in January or February.

In 1861 the system was changed so that the miners were paid twice a year, in May and November.[7]

The establishment of a regular month for payment was aided by the gradual evolution of more efficient accounting systems. The main cause, however, was the increased ability of the mining concern to raise money for the pay, even when lead prices were low and sales few. The early Blackett/Beaumont records, particularly the chief agents' letter books, have many references showing the difficulties of raising money and the reactions of the workmen to delays in holding the pays.

In October 1730, the chief agent wrote that 'Mr Peart [the Weardale agent] tells me, the miners in Weardale are grown so refractory for want of a pay that he is sure the greatest number of them will throw up if not pd. before Xmas: where money will arise to pay the Interest in arrear, make Weardale pay, and pay Mr Mowbray's ballance, I know not'. In December he received a petition from the Weardale miners asking for a pay at once, 'and since the beginning of last Month have been oblig'd to let Mr Peart have £1200 to make them easy, without which I am assured not half of them would have taken on his bargain now at Xmas: there is still owing to these groves about £4500 to Michas. last'. In April of the following year he wrote, 'I have been almost pulled in pieces in Weardale' by the dissatisfied miners. They were apparently paid off some time during that year, 1731. In December 1732 matters had improved so greatly that the chief agent could write to Sir Walter Blackett, 'After Weardale pay is made you will have paid off all your works this year, both mines and mills, which has not been done for 20 before'.[8]

At this period—the 1730's—the workmen in the different mining districts and those in the mills (including the carriers) were paid off separately as money arose. By about 1760 the pays of all the different districts took place on successive days in the same month.

In 1758–62, when lead prices were very low, there is another spate of references to the difficulty of raising money for the pays. In September 1758 the chief agent wrote to Sir Walter Blackett that it was impossible to borrow money on the security of unsold lead either in Newcastle or in London. 'I therefore see no other way than to let the people have what subsistence money I can raise to the spring, in hopes of getting a quantity of lead sold against that time.'[9] In March of the following year he wrote, 'I had some more of the Weardale people and some of the lead mills pressing for their respective pays, for want of which they alledge their landlds. threaten to break them up. I put them off in the best manner I can for the present and hope to be able to satisfy them all before Mayday if you can help me with £5000.'[10] The Weardale pay was made during May; the Allenheads and Coalcleugh pays, where Sir Walter himself was the chief landlord, were left over until later in the year. In November, however, the agent wrote that he was forced to sell some lead at considerable loss to pay off Allenheads and Coalcleugh, 'for the people here can't want the Money, but as far as the Workmen in Weardale they are in better circumstances and may be put off the next year with a little Subsistence.'[11]

One particular unfortunate body of miners were those employed in the 'partnership mines' in Weardale, leased jointly by Sir Walter and Mr Bacon. These could not be paid until both parties were ready, and there are frequent references to failures to pay up. In 1761, 'As for the ptnership miners, there are 2 years due, which looks badly, but if I had the money I could not pay them without Mr Bacon, who I do not find makes any preparation towards it.'[12]

The sums of money required to make the pays were considerable. The Blackett/Beaumont accounts do not record wage payments as such, but in the letter books are occasional references to sums paid in wages between 1760 and 1811. The figures are not strictly comparable since they do not refer to identical periods of time (these not being stated). Nevertheless they are quoted here since they give a clear impression of the order of magnitude of the wage bill and its growth during the eighteenth century:

1760	£14,000 for all the mines but *not* the mills.
1761	£6,000 for Weardale + £2,000 for the 'Partnership' mines.
1767	£17,290 for all the mines.
1779	£7,000 for mills and carriers.
1781	£25,498.
1784	£36,000.
1786	£41,000.
1787	£43,722.
1788	£51,613.
1793	£45,000.
1794	£65,550.
1796	£56,000.
1797	£53,000.
1799	£51,000.
1811	£68,622.

By the end of the eighteenth century, in spite of the increasing size of the amounts of money required to make the pays, a more regular annual date of payment had been achieved. References to the pays in the letter books become less common, showing that the problem of raising money had lessened greatly. In 1812, so accustomed had the men become to regular payment that a delegation from Coalcleugh protested when it was delayed a couple of months, from March to May. This would have been nothing to the miners of the 1730's and '50's.

TABLE I

FREQUENCY AND AMOUNT OF SUBSISTENCE PAYMENTS,
LONDON LEAD COMPANY AND BLACKETT/BEAUMONTS, 1766–1875

	Blackett/Beaumonts		London Lead Company	
Year	*Frequency of payments*	*Amount paid (per month)*	*Frequency of payments*	*Amount paid (per month)*
1766	Bi-monthly			
1778–84	Bi-monthly			
1785			Monthly	
1790–94	Bi-monthly	10s 6d		
1795	Bi-monthly	10s 6d	Monthly	10s 6d
1796	Bi-monthly	15s 9d		
1797	Bi-monthly	Increased to 26s 3d		
1798–1800	Bi-monthly	31s 6d		
1801–12	Monthly	30s		
1816	Monthly	Reduced temporarily to 20s	Monthly	Increased from 30s to 40s
1818	Monthly	30s		
1833	Monthly	30s	Monthly	40s
1842	Monthly	30s	Monthly	40s
1845	Monthly	30s		
1846	Monthly	Increased to 40s	Monthly	40s
1854	Monthly	40s	Monthly	Increased to 44s
1862–c.1875	Monthly	40s	Monthly	44s
1872			Monthly	Increased from 51s 6d to 54s
c. 1875	Introduction of 'new system'.			

Later in the nineteenth century there are no references at all to delays in payment, so presumably there were none.

Unfortunately there is no similar historical record of the payment practice of any other lead mining company. For the London Lead Company there is only a court minute of 1785 recording a change of practice—'That the Company's pays in the North which used to be paid Half-yearly be in future paid only once every Year up to Michaelmas 1785 and so to be continued.'[13] In 1833 the Beaumont chief agent wrote of the company that 'They make their great Pay in January, we in March or April.'[14] The Beaumonts too were making their annual pay in January in the 1840's.

The Blackett/Beaumont records show, therefore, that in the early and mid-eighteenth century pays were infrequent and irregular; by the end of the century they were held regularly once a year. From 1861 onwards the men were paid every six months. The records of payment of subsistence between pays show a similar pattern. Early in the eighteenth century, payment was irregular, and of no fixed amount. By the end of the century bi-monthly payments of a fixed sum were usual. From 1801 onwards the subsistence was paid monthly. Table 1 illustrates the changes in tabular form, so far as evidence is available. There is rather more about the London Lead Company than there is in the case of the pays; its employees received monthly subsistence from at least 1785 onwards.

Subsistence money

The letter books in the Blackett/Beaumont records show the irregularity of subsistence payment in the early eighteenth century. Money was as difficult to raise for this purpose when lead was selling badly as it was to raise for the actual pays. In 1759, for example, the agent did 'not know what can be done for money this year for the peoples' Subsistence'.[15] In theory it was paid every two months by 1766.[16]

Table 2 is based on the only surviving account books recording the dates on which subsistence was paid, and the sums involved—for Weardale from 1790 to 1812. From 1790 to 1800 the money was paid roughly every two months. No definite day was set, however, and there were sometimes gaps of as much as three months or as little as one between payments. After 1801 there was a regular day for payment every month. Monthly subsistence remained the practice from then until the Beaumonts abandoned mining in the 1880's.

The amount of subsistence paid to individual miners in the earlier part of the eighteenth century was also irregular; the money was divided up among the men as it became available. There is no concrete evidence of sums paid until the end of the century. By 1790 each adult miner employed by the Beaumonts received £1 1s every two months. A London Lead Company miner received the same amount, but it was paid monthly, 10s 6d at a time. In 1796, after a strike in Weardale, the Beaumont chief agent 'thought it advisable *at this time* to increase

TABLE 2

SUBSISTENCE PAYMENTS TO INDIVIDUAL MINERS IN THE BEAUMONT
MINES IN WEARDALE, 1790–1812

Year	Number of payments in the year	Sum paid on each occasion
1790	6	£1 1s
1791	7	£1 1s
1792	6	£1 1s
1793	7	£1 1s
1794	6	£1 1s
1795	5	£1 1s
1796	7	£1 11s 6d
1797	6	Two of £1 11s 6d
		Two of £2 2s
		Two of £2 12s 6d
1798	7	Three of £2 12s 6d
		Four of £3 3s
1799	6	£3 3s
1800	6	£3 3s

From July 1801 to 1812 regular monthly payments of £1 10s were made.

their subsistence Money one half', to £1 11s 6d every two months.[17] The miners were still not satisfied. One of the requests in the Weardale petition of 1796 was that they should receive 'one guinea per month each man'. The Weardale accounts of subsistence show the great increase in payments that took place between 1795 and 1801. A miner working in 1801 received more than double the sum he received in 1795: 30s every month, instead of 21s every two months.

Thirty shillings remained the standard monthly subsistence until it was raised to 40s in 1846. In 1816, a year of severe depression in the lead trade, it was temporarily reduced to 20s[18] but was back to 30s again the following year. In 1818 there was another strike in Weardale. In a petition dated 2 October 1818 the strikers wrote: 'We find that our present subsisting Money which is only 7/6 a Week, much too little to purchase the necessaries of life. We have therefore petitioned for 10/- a Week. Our requests have not yet been granted. . . . Unless the Subsisting Money is advanced to 10/- a Week, it will be impossible for the greatest number of us to get the necessaries of life, as our Credit is utterly gone' (see Appendix 4). The earlier request had been rejected by the chief agent—'a demand so unreasonable cannot be complied with'[19]—in spite of the fact that the London Lead Company was paying its miners 40s a month. When the strikers went back to work they still received only 30s.

A few scattered figures are all that survive in the London Lead Company records to illustrate changes in subsistence rates, and they show none of the reasons behind the changes. In midsummer 1816 the rate was the same as the Beaumonts', 30s a month. At the end of that year the new chief agent, Robert Stagg, raised it to 40s a month. In 1833 this sum was payable only 'to Pickmen in Regular divisions and places . . . so long as they use due diligence and conduct themselves satisfactorily. The advances for old Pickings to be confined to the actual Earnings, as near as the same can be ascertained, as heretofore'.[20] This was a severe limitation, as a large proportion of the labour force was employed in the 'pickings' in this period of bad depression, when lead prices were too low to encourage new exploration; many miners were unable to earn enough to keep themselves and were forced to leave the region.

Why did the subsistence system exist at all? Before outlining the changes that occurred in the middle and later years of the nineteenth century it is necessary to consider its appeal to both employers and workers in the lead industry.

From the employers' point of view, the system of annual pays with subsistence in between allowed them time to raise the large amounts of money necessary. There was a lengthy interval between the extraction of the ore from the ground and its sale as lead. When demand was low, it was better to stockpile for as long as possible rather than sell at an uneconomic price. Irregular pays, as and when cash was available, therefore suited the eighteenth-century mine owner.

For the miners, the system was a guarantee that they would receive some money during the year, even if luck was poor and hardly any ore raised. A partnership might get hardly any ore for three of the four quarters yet strike lucky during the fourth. Alternatively, a man might have no success for two or three years and end up with a debt instead of a pay. But he would continue to receive his subsistence until, with luck, he had a successful year and was able to repay the debt.

The owners often claimed that the miners also benefited from a system that prevented them from having too much money in their hands at one time, and gave them an extra annual sum to meet debts with. In 1766 the Blacketts' chief agent wrote to Sir Walter Blackett that 'two months subsistence at a time proves a temptation to extravagence'[21] but that the system should nevertheless be continued. In December 1795, when the subsistence money for the Beaumont miners had just been increased from 21s to 31s 6d per two months, the Weardale miners asked for 42s. The chief agent wrote to T. R. Beaumont that the demand 'would on *no Account* be prudent to comply with. Were you to do it their families would not be benefited by it and at the time of the pay the greater part of them would have little or nothing to receive, their creditors would be clamorous, and you would incur blame for the badness of the pay, as they term it, when they have little to receive.'[22]

A Beaumont agent said much the same in 1842 when the commissioners questioned him about the difference between the London Lead Company's payments to its workmen and the Beaumonts' (30s per month). 'The Lead Company advances 40/– a month, which is good for the men at the time, but when the men reckon at the end of the year it is usually found that what has been advanced to them comes to as much as what they have earned, and they have nothing to receive. It is beneficial for the miner to have to receive a sum of money, as he can then provide himself with necessaries, which he would not have been able to save sufficient money for if on weekly wages.'[23] R. W. Bainbridge, the Lead Company's chief agent, told the 1864 commission that the annual pays were useful to miners to provide money for cottages and stock for their smallholdings. 'It comes to meet those annually accruing charges which they cannot escape, and our retaining the money in our own hands I believe is better than if they had it in their pockets, in many cases.'[24] The unexpressed reason why the companies did not wish to raise subsistence payments too much was that they did not wish to risk too large a proportion of men being in debt at the end of the year.

By the nineteenth century the system was under pressure from both sides. The mine owners were able to make use of the greater fluidity of the money market to borrow money when necessary to overcome temporary difficulties. They were able to establish regular times for payment by the end of the eighteenth century. A regular monthly wage, disguised as subsistence, gave the mine owners more extensive control over the miners than irregular pays and low subsistence. Both the big companies were willing to increase subsistence money greatly, and even to write off part of any debt owing to them at the end of the mining year, in return for the greater power over their workers.

The miners themselves were always pressing for more. In the petitions from Weardale miners (1796 and 1818) higher subsistence payments were among the most prominent demands. The disadvantage from their point of view was that it ate into the amount they received at the annual pay, and might absorb it completely. Assistant Commissioner Mitchell put the position well in 1842:

But it may happen that, instead of having money to receive, the miner is found not to have earned as much as stands in the book against him. In that case the balance is struck, and entered against him in his next year's account, and it may go on year after year; but if he shall in course of time have a good year he may be able to extinguish all his debt, and even have something coming to him. It is very disheartening for a miner to go on year after year and have nothing to receive at the year's end; but still he has his lent-money on the first Friday of every month, and he lives in hope.[25]

In some ways the possibility of large debts after a bad year was helpful in bargaining with the masters. The latter did not want the miners to be in debt any more than they themselves did. In 1811 the Beaumont chief agent wanted to increase the bargain prices, in spite of the depressed state of the lead trade, as they were 'let lower than they can be supported for twelve Months together, without disabling the Men to meet their advance Cash, and if so, will increase the Mining Debt.'[26]

There are a few scattered figures in one of the Coalcleugh account books, jotted down by the agent on the reverse of the accounts proper, of debts owing in the period 1824–35. The total outstanding in 1824 was £182 1s 7¼d. The oldest dated from 1806, the rest from 1813: the highest sum involved was £10 10s 6d and the lowest 14s 6d; the total number of men owing money was thirty-two. During the bad depression of 1831, the total sum amounted to £487 4s 6d. One hundred and twenty-three men owed money, the most being £15. At least £180 16s 2½d of this was later deleted from the accounts as 'in all probability will never be got'. By 1836, when the lead trade was more prosperous, only seven miners owed money.[27]

Thus in normal times the amount of money owed to the mining companies was small in relation to the labour force and the enormous sums paid out in pay and subsistence. In periods of severe depression the debts inevitably rose, and a large proportion had to be written off. The companies were therefore in a position to be generous to unlucky miners.

To return now to the account of the development of subsistence payments, the London Lead Company kept 40s as the usual sum payable for nearly forty years. In 1856 it was increased to 44s per month. According to the chief agent in 1864, this was the sum paid to an adult miner. New members of a partnership were paid 35s a month the first quarter, and then 40s a month until they were 21.[28] By 1872 the monthly sum paid had become 51s 6d; in that year it was raised further to 54s.[29] A large proportion of their total earnings, therefore, was being paid to the miners in the form of subsistence. This was almost certainly linked with the Lead Company's practice, by the middle of the nineteenth century, of employing nearly all its miners on a fathom instead of a bing basis. By this means the earnings of the miner could be much more easily controlled and the price per fathom, related to the hardness of the ground, judged so as to allow the partnership members to earn their subsistence money and a little over.

The Beaumonts continued the old practice of extracting ore by the bing. This was less efficient in that it was impossible to regulate the earnings of the miners so closely as when the fathom system was employed, but it did spur them on in the everlasting hope of striking a really rich vein. To make up for his relatively low subsistence payment, the Beaumont miner could expect a larger pay than a Lead Company miner, and in addition after 1846 a guaranteed cancellation of half of any debt incurred during a single year.

When Sopwith became chief agent in 1845 he was immediately petitioned for a larger subsistence payment.[30] The following year he increased the payment from 30s to 40s a month and introduced a number of new conditions and provisions at the same time. The workmen of the different districts were told of the changes at meetings, and Sopwith published, for free distribution to them, a pamphlet entitled *Observations addressed to the miners and other workmen employed in Mr. Beaumont's lead mines . . . 2 February 1846.*

This pamphlet records, in Sopwith's usual verbose style, a number of important innovations. First, the monthly subsistence was to be raised to 40s. Second, an elaborate scheme was introduced to cut by half the amount of money an unlucky miner might owe at the end of a year. Thus if a miner finished the year owing £5, only £2 10s was transferred to the following year's books. To pay off the other half of the debt, every miner whose average earnings throughout the year exceeded 10s a week was to contribute 10 per cent of the excess money earned. If his earnings were 25s a week he would contribute 52s, or 1s a week. This sum was, of course, deducted from the money he was to receive at the pay, not from his monthly subsistence. If half the total debt was not met by this levy on the successful miners it would still be cancelled, and the Beaumont concern would bear the loss. All debts outstanding in February 1846 would be cancelled.[31]

The miners welcomed this part of the scheme. Sopwith's additional condition 'that the advance of forty shillings has reference to actual work performed during five days of eight hours each; that is to say, forty hours per week . . . which at three pence an hour, amounts to forty shillings per month' was the root cause of the unsuccessful Allenheads strike of 1849.

The same subsistence was paid to the Beaumont miners until the mid-1870's after Sopwith's departure. In 1875 some of the bargains taken out in the Weardale district concluded with the words 'Money to be advanced according to Quantity of Ore raised.' There is no information in the Beaumont records themselves as to how this was interpreted. All that exists is a reference in a pamphlet issued during the last of a series of bitter strikes in Weardale in 1882, coinciding with the Beaumonts' withdrawal from lead mining.[32] In this pamphlet a lengthy contrast is made between the 'old system' of paying subsistence (Sopwith's), and the new system introduced by his successor as chief agent, J. C. Cain.

The 'New System' (or 'Cain's System' as it is often termed) is very intelligible. Any miner working under it gets, or is supposed to get, 80 per cent of his earnings for his monthly subsistence-money, and the remainder at the half-year's end; so that you see that the poor miner's subsistence money is regulated by what he makes, if he is so unfortunate as not to make anything during the month, there is no subsistence money for him at the month's end, although he may have worked very hard all the month.

The late 1870's were a time of rapidly falling lead prices; the 'new system' was introduced to save the Beaumonts from losses. For the lead miner, with his uncertain and irregular successes, it was utterly ruinous.

Nothing has been said about payment and subsistence in the smaller mines because no evidence exists save the few figures given in the reports of 1842 and 1864. Quarterly settlement was apparently usual at more than one Alston Moor mine.[33] At the Derwent mines, where the Cornish system of bidding for sets was in operation, the miners were given 7s a week subsistence in 1864.[34]

The method of payment

The annual pay was the greatest festivity of the miner's year, at least in the eighteenth century and the early part of the nineteenth. The Blackett/Beaumont miners were paid in their respective districts on successive days. In the eighteenth-century letter books there are a number of letters from the chief agent in Newcastle to the district agents, instructing them as to procedure. A typical one regarding the pays in 1778 runs:

Yourself accompanied by the rest of the stewards to be here on Friday Evening the 2nd of January—receive your money on Saturday the 3rd—go away on Sunday the 4th and make Dukesfield pay on Monday the 5th—Allenheads on Wednesday the 7th, Coalcleugh on Thursday the 8th and Newhouse pay on Friday the 9th . . . . As these sums will amount to a Considerable Sum of Money you are desired to come well armed.[35]

The payment of the Weardale miners at Newhouse, the home of the resident agent, is very well described by James Losh, who attended as auditor in 1828.

Newhouse (Mr. Crawhall) is large, old and inconvenient. The entrance hall is a long narrow room with a table the whole length of it, at which the *pays* are made. Mr. Crawhall sat at one end of this with one plate full of sovereigns, another of *silver* and a third of copper coin before him, with piles of bank notes (the large ones Batsons, the small Scotch) on one side of him under the care of a clerk. Three other agents or clerks assisted in keeping the cheque accounts so as almost to prevent the possibility of any mistake. The workmen were admitted in regular order and received their balances, upon respectively producing a ticket, shewing what was due upon work done under the original bargain, deducting what had been received for *subsistence*, etc. . . . . Near Mr. Crawhall's house there were about 40 tents pitched, many of them supplied with liquor and refreshments, cold meat etc, but many also containing Yorkshire cloth, hats, shoes, trinklets, etc. for sale. A curious example how closely supply follows demand; how soon money makes a market. Many shopkeepers too from Newcastle who supply the retail dealers in this district with groceries, hardware, etc. were in attendance to have their bills discharged. The *pay* was made to the workmen: they paid their bills to shopkeepers of this district and they again the persons of whom they made their wholesale purchases. It is said that Mr. Featherstone (a grocer in N. Castle) generally receives during the pay about £8000 . . . . About 20 of the principal agents, etc. dined with Mr. and Mrs. Crawhall and myself and I suppose not fewer than 100 of the inferior agents, farmers, etc. in the kitchen. Mr. Bolam, Mrs Beaumont's Land Steward, attended to receive the farm rents which also mostly are obtained from the *pay* at first or second hand.[36]

When Thomas Sopwith became chief agent in 1845 similar ceremony was maintained. In 1847 he received the money for the pay 'With about 10 of the Smelting and Mining Agents who came down to Newcastle in conformity with an old custom when the object was to carry the money on horseback and secure safety by a numerous cavalcade.'[37] Writing in 1862 he recalled the details of the pays when he first became agent, 'of crowds of people—of rows of booths and tents—of gay and attractive performances by travelling actors—the banners flying and the band playing the White Cockade as the Money for the pays entered the village' [Allenheads].[38] But this had become a thing of the past. Sopwith disapproved of the interruption to work, the drinking, the waste of money on inessentials. In 1858 there were very few tents. Sopwith had 'advised' the workmen not to assist in erecting them.[39] In 1861 'There are no booths in the village. . . . Entire amount paid in less than two hours.'[40]

In Sopwith's view, subsistence payments were similarly badly organised. In September 1845, 'owing to the want of methodical arrangement nearly the whole of the afternoon was occupied in making this advance. A number of men were thus kept waiting three or four hours with the only alternative of sauntering in the open air or of taking refuge from cold or inclement weather in the Public House.' By March 1847 the 'advance was made on the same day [as the money arrived from the bank] in less than an hour'.[41]

It was common in some industries for the masters to pay their employees in groups with coins or notes of high denomination only. These could be changed at a shop or inn owned by the masters, and much was spent on drink. Writing in the 1870's a Newcastle journalist described a pay at one of the small, independent mines on the slopes of Crossfell on Alston Moor. 'They work in gangs of six or eight, and are

F

paid quarterly. One of the party is made cashier, and he receives the money for them all. On Receipt of payment they all adjourn to the nearest public house, when the cashier divides the money by handing each man a note, a sovereign, and so on through all the coin. Whatever odd money remains is spent in drink.'[42] This was probably the method of payment employed by most of the smaller lead mining concerns throughout the eighteenth and nineteenth centuries.

Neither of the two largest companies appears to have been guilty of this practice—at least from the end of the eighteenth century onwards. Both paid their men at their respective company offices, and each miner was paid individually.

The London Lead Company codified these points in a court resolution of 6 October 1785: 'That the monthly advance money & also what becomes due to them be paid to them themselves and no other person, in money and not in any kind of Bank Notes, nor any Trades people attend at the pay: nor to any order written or otherwise, except the Miner is ill & cannot attend himself.'[43]

The Blackett/Beaumonts did not work by any such rigid resolutions, but it is apparent from the correspondence and accounts that their practice was much the same. Before every pay in the eighteenth century there was a desperate scramble for money—and it had to be the right kind of money, difficult to raise in the days before an adequate local banking system had developed. Banknotes were used, but low denominations were preferred. It was always a problem getting sufficient cash. In 1765, for example, Sir Walter Blackett's land agent in Hexham was asked to supply as 'much silver as you can muster'.[44] In 1792 the chief agent wrote that 'neither the lead pays nor the Subsistence for the Workmen could be made in Bank of England Notes without a very considerable Sum in Cash; we have found that one third in Cash is required at the Lead Pays; and one half for the Subsistence'.[45] For the lead pay of 1797, amounting to £53,000, the chief agent was able to obtain £37,000 in notes of £5 each, £14,000 in notes of £1 each, and £2,000 in gold and silver. 'This was a totally inadequate proportion of cash and I much fear that it may not give satisfaction.'[46] But the Blackett/Beaumont accounts show that the concern was as rigid as the Lead Company in paying each miner individually. In the subsistence accounts for the period 1790–1812 only members of the same family, fathers and sons, were paid jointly.

The miners' earnings

The system of payment and the inadequacy of the surviving accounts do not allow a very precise historical narration of the miners' earnings. It is possible to give a general account of rises and falls in income over a number of years, but precise figures to compare with known wages in other industries are unobtainable. We have already seen that subsistence payments increased greatly from the end of the eighteenth century onwards. These payments must then have formed the major part of the income of all but the most fortunate miners. In Featherston's book about Weardale, published in 1840, there are a number of 'reconstructions' of miners' conversations. One miner remarks to another, 'I'm certain we have not cleared our powder and candles, besides sharpening and drawing; but we mun be content; it's better than being off work: ne lent-money coming in does not answer.'[47] This must have been a common attitude. Dr Mitchell, who observed that the miners always lived in hope, also observed how seldom the hope was realised:

Taking refuge one day from a heavy storm of rain, near the head of Weardale, I found the parlour of the inn full of persons connected with mines. I turned the conversation so as to inquire in what condition the working miners in that part were. An overlooker of work in the mines said to me 'Sir, the miners are exceedingly well off here. They earn a great deal of money. I could point out six men who each of them have made up £60 at the year's end, and they have done so for these nine years past.' I replied 'I am very glad that these men have done so, but I think I should have no difficulty in finding you 600 who have not earned half the money.' The company all said I was right, and the man himself admitted it; yet he would have passed his rare case as an ordinary sample if he could. Such a manner of acting is but too common with some masters, and the persons whom they put in office.[48]

The accounts of the Blackett/Beaumont mines give only some indication of the miners' general prosperity; there are no specific figures of annual earnings. The surviving accounts are the general accounts of the costs and production at each mine. No wage books or their equivalents survive. Thus it is recorded that '— and partners' raised so much ore at such and such a price during a quarter, but the partnership's expenses—washing, drawing, candles, etc.—are not recorded, and the same partnership may appear elsewhere in the accounts for specific dead work or days' labour performed. It is thus hopeless to attempt to work out from these figures the actual year's earnings of an individual partnership.

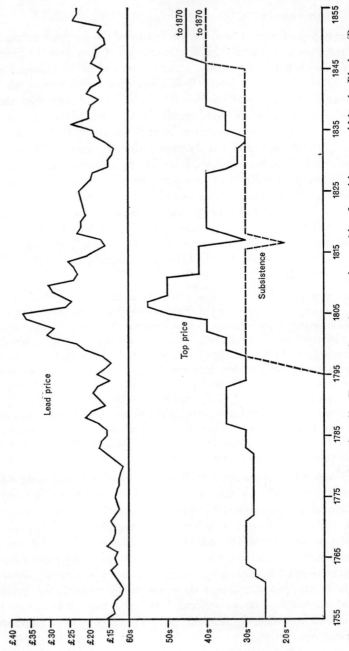

Fig. 2 Graph comparing the market price of lead, 1755–1855, the top price per bing for raising ore paid by the Blackett/Beaumonts, 1755–1855, and the subsistence rate per month paid by the Blackett/Beaumonts, 1790–1855

It is, however, possible to obtain a picture of overlying trends—the general level of prosperity during different years, not allowing for individual cases. In Fig. 2, three sets of figures are displayed. The first curve represents the average sale price of lead per ton during every year from 1755 to 1828 as obtained by the Blackett/Beaumonts, and from 1829 to 1855 as given by Tooke.[49] The price of lead, like that of most other metalliferous products, fluctuated wildly from year to year, or even from season to season. War or the threat of war sent the price racing up; with peace, it came down again. The connection between the market price of lead and the earnings of the miners is obvious. If prices were low, the owners could not afford to encourage their workmen with high bargain prices; if they were high, bargain prices were also high to encourage the men to work as hard and fast as possible before prices fell again.

The second curve is based on the highest price given per bing of ore at the Blackett/Beaumont Coalcleugh mines during the years 1755–1870. As has already been explained, the price per bing of ore was regulated according to the richness of the vein and the difficulty of the ground. Throughout this period, however, the individual mine agents were bound by a 'top price', above which they were not to go, however poor the ground. This top price was altered according to the selling price of lead. At any one bargain letting there were normally more miners contracting on the basis of the top price than on any other. For example, at Coalcleugh at Michaelmas 1771 there were twenty-five partnerships employed at between 15s and 26s a bing, and fifty-three at the top price, 28s. At Michaelmas 1792 there were thirteen partnerships employed at between 18s and 28s, forty-one at 30s, and forty-six at the top price, 35s. Hence, the top price is a meaningful, if crude, indication of the miner's relative prosperity.

The third curve is based on the monthly subsistence payments to the Blackett/Beaumont miners. This again bears a close relationship to the overall economic situation of the mines as dictated by the market price of lead. (There is no record that subsistence payments were reduced during the early 1830's; quite possibly they were.)

Two other, less reliable, indicators of comparative earnings are given in Table 3. These are, firstly, estimates made by various observers at different times of 'average' annual earnings. Unfortunately these figures are little better than guesses, based on the experiences of a single miner or a very small sample. Also shown in the table are the intentions expressed by the mine owners at different times as to what should be the earnings of their miners. Again, the bargain system did not allow these intentions to be translated easily into practice.

TABLE 3

CONTEMPORARY ESTIMATES OF THE ANNUAL EARNINGS OF
LEAD MINERS

Year	*Estimate*
1765	Between £48 and £160 (almost certainly a very excessive estimate).[1]
1797	'From £5 or £6 to £50 or £60 a year; the average is about £25.'[2]
1800	London Lead Company 'basis' raised from £26 to £31 4s.[3]
1802	London Lead Company 'basis' £31 4s. [3]
1805	London Lead Company 'basis' £39.[3]
1815	London Lead Company 'basis' £30 to £35, 'which is really less than they can live on.'[3]
1815	London Lead Company 'basis' £31 4s.[3]
1816	London Lead Company. Would make £31 4s. if they worked a six-day instead of the customary 5-day week.[3]
1816	Beaumont miners earning between £15 12s and £31 4s.[4]
1822	A Greenwich Hospital report estimated that pickmen on Alston Moor earned between £39 and £52.[5]
1828	Allenheads miners average £39.[6]
1831	London Lead Company miners averaging under £26.[3]
1832	Weardale miners in the worst places earning less than £18 4s.[4]
1834	Miners in Alston parish earning between £18 4s and £20 16s.[7]
1836	London Lead Company 'basis' £32 10s.[3]
1842	London Lead Company estimate between £31 4s and £39.[8]
1842	Beaumont estimate between £37 14s and £39 17s.[8]
1842	Derwent company estimate between £23 8s and £26.[8]
1846	Beaumont 'basis' £39.[9]
1848	London Lead Company 'basis' £31 4s.[3]
1854	London Lead Company 'basis' £39.[3]
1864	Beaumont estimate from £39 to £44 4s.[10]
1864	Rodderup Fell Company estimate from £40 to £50.[10]
1864	London Lead Company 'basis' £39.[10]
1874	Derwent Mining Company actual average earnings during July (from a calculation in a report book) 21s 6d, i.e. £55 18s per year.[11]
1878	London Lead Company 'basis' *reduced* to £41 12s.[3]

Sources

[1] G. Jars, *Voyages métallurgiques*, 1774–81.
[2] *Sir* F. Eden, *State of the poor*, 1797.

[3] London Lead Company records.
[4] Blackett/Beaumont records.
[5] Greenwich Hospital records.
[6] J. Losh, *Diaries and correspondence*, 1959.
[7] Poor Law Commission report, 1834.
[8] 1842 report.
[9] T. Sopwith, *Observations addressed to the miners*, 1846.
[10] 1864 report.
[11] Derwent Mining Company records.

Note: each figure is almost meaningless by itself, as the 'average lead miner' did not exist, but the estimates are interesting as showing the general trend of earnings. All the estimates—originally expressed in daily, weekly, monthly, etc., terms—are here expressed in yearly terms

From the statistical and other evidence, certain factors may be isolated as vitally affecting bargain prices and the miners' eventual earnings. The fundamental importance of the market price of lead has already been emphasised. Jars, who visited the mines when the market was buoyant, wrote that this sytem, 'quoique profitable aux intéressés, tend absolument au désaventage de l'exploitation'. Dead work, necessary for long-term exploitation, was neglected to obtain a quick profit.

De ces prix-faits mal entendus, il resulte que les salaries des ouvriers sont fort souvent beaucoup plus considérables qu'il ne conviendroit; chacun d'eux gagne quelquefois dans trois mois de travail, depuis 12 jusqu'a 40 livres sterlings, ce qui est exorbitant. Les intéressés disent à cela qu'il leur est absolument égal, parce qu'ils ont fait leur calcul de manière que le produit du minérai est toujours de 50 pour cent, même au-dessus et jamais au-dessous, et que d'ailleurs cela fert d'encouragement; cela peut être vrai dans un sens, mais en général un ouvrier qui gagne trop est rarement bon ouvrier.[50]

Jars's opinion of the deleterious effect of higher bargain prices (his figures for annual earnings, though, are suspiciously high) were shared by the mining companies at times of a booming lead market. A London Lead Company agent wrote in 1808, just after a great boom in prices, 'Since the advances in the Lead Markets many of the Mining Companies in this country have advanced the price for raising ore in their Old Works to 70/– and some to 80/– per Bing which makes our way with the workmen very difficult.'[51]

Such times of boom were rare. When lead prices fell, the companies had the choice of three courses of action, all vitally affecting their employees. Prices could be reduced, sometimes to starvation level. In 1815 the Lead Company's chief agent wrote that 'the prices were kept so low during the three last Quarters that the Workmen's earnings were quite inadequate to the support of their families'.[52] They were to fall even lower the following year.

The second step was to reduce subsistence payments. In 1816 the Beaumonts lowered the monthly subsistence rate from 30s to 20s. Around 1830 the London Lead Company paid a regular subsistence sum only to men in good places. Those working in the 'pickings' merely received the estimated value of the ore they raised.

The third step, which will be considered further in chapter IX, was to cut down the labour force. If the depression were not too prolonged, the men would remain in the area unemployed until trade revived and employment once again expanded. In a severe and long lasting depression, as in the early 1830's, many families were forced to migrate. The labour force was cut back either by directly refusing to employ men, or by offering them a price too low to afford a living wage. At the small mines of Heartlycleugh and Keirsley Row, during the depression of 1831, the Beaumonts' Coalcleugh agent reported 'Have not been able to let a bargain at either of those Mines', the prices offered being so low.[53] Many of the smaller mines on Alston Moor were forced to close down completely during slumps as they had insufficient capital to tide them over until trade improved.

The mines of the larger concerns were saved from closure by the need to keep some labour together for the arrival of better times, and by the desirability of keeping maintenance and dead work going. It was, of course, cheaper to carry out dead work in such times—provided the company concerned had enough capital to do so—and lead which would only be sold below cost was better left in the ground for the time being. The court of the London Lead Company recorded in 1829 that 'however deeply they may be impressed in the necessity of practising the strictest economy in all branches of the Company's expenditure, cannot but concur with Mr. Stagg [the chief agent] in the propriety of continuing the different trials, now in progress, to their ultimate object, believing that it is necessary for the future prosperity of the Company to have numerous Mines ready for working when required, particularly as such work can never be so advantageously performed as at a period like the present when the price of labour is so materially reduced'.[54] Even the London Lead Company could not continue this policy for the whole of the depression of the late 1820's and early 1830's; dead work was almost halted before it was over.

Another important trend, frequently commented on by the mining companies, was the gradually increasing expectations of the miners, and their reluctance to give up increases in bargain prices when lead prices went down. A special committee of the court of the London company reported on bargain price policy in 1793:

In the Year 1788 (very unexpectedly) Lead rose to the astonishing price of £23 per fother, then the miners took advantages and rose the Bargains to 35/–, 38/–, and 40/– per Bing in the Years 1789 and 1790. The Lead Market had considerably fallen from £17 to £16 and tho' your Agents had not Lowered Bargains in proportion from 35/– to 30/–. Thomas Dodd [the chief agent] has had some consideration on the fallen state of the Lead Market by lowering the Bargains since Ladyday 1793 about 3/– per Bing. He has not nearly reduced the Bargains to the same as they were in the Years 1785 and 1786 when Lead sold at the same prices nearly as this year—£16-10. It may be justly observed by the Court *that this ought to be done* but we fear there will be some difficulty in accomplishing that object, tho' it is the duty of the Court to attempt it, but it is worth considering whether the Company shall stain it a shadow upon the eve of perhaps a severe winter.[55]

In a joint 'memorial to government' asking for a reduction in the export duty on lead, all the northern mine owners stated that before 1793 'the wages of the Working Miner were 1/2 per day'. But in 1814 their wages, 'tho' in some degree reduced since the Peace, are now 2/6 per day', with only a small increase in the price of lead compared with the earlier date.[56]

An interesting but unfortunately brief note in Thomas Sopwith's diary for 23 June 1866 complains about an unexpected result of the building of the new railway from Hexham to Allendale town. Conversations with the navvies were spreading dissatisfaction among the miners. They were learning that higher wages could be earned elsewhere, and Sopwith feared that the railway would facilitate migration.

During the nineteenth century both the major lead mining concerns attempted to maintain a policy of steady bargain prices, irrespective of fluctuations in the price of lead. From 1846 to 1870 the Beaumonts managed to hold a steady top price and a regular subsistence rate. Sopwith's system for equalising earnings has already been described. The London Lead Company pursued a less subtle but probably more effective course. As early as the end of the eighteenth century, the management attempted to base all bargain prices upon a theoretical weekly wage which all the miners should attain but not exceed. Thus if 12s a week were the agreed figure, all bargains would be calculated to produce just that and no more. In 1800 the company's chief agent 'assured the Court that the price of Lead had no influence upon the price of Bargains.'[57]

The fragmentary figures of bargain prices reported to the court in the period 1801–15 show this statement to have been untrue. In 1801 the price of lead was around £25 a ton; in that year the *average* Lead Company bargain price was between 33s 6d and 36s a bing. In 1806, when lead was over £35, the average Company price was between 39s and 47s 6d. In 1811 lead was back to around £25; bargain prices were down to an average of 32s by the end of the year.

Nevertheless the company's court minutes show that bargain price increases were reported to it as consequent upon the increase of food prices rather than a rise in the price of lead. In 1800 'the Bargains were let at present so as to produce the Men 12/– per Week in consequence of the high price of provisions instead of 10/– which is the customary Wages'.[58] Bargain prices were reported to the court as the 'wages' which they would produce. At this time—before about 1817—such a wage figure must have been fiction. It was virtually impossible, bearing in mind the complexity of estimating the productivity and difficulty of a particular place within a mine, to engineer prices so that all partnerships would earn the same.

After 1817 the figures become more meaningful. At that time Stagg introduced the 'new fathom' system under which men were paid for the amount of ground they cut away, regardless of its ore content. Stagg particularly claimed as an advantage of this system that 'the Wages will be more equalised' because it was relatively easy to calculate the work that could be done in a quarter and to regulate the price so as to produce standardised earnings.[59]

Apart from the slump of the early 1830's, the London Lead Company maintained a steady wages policy from 1817 on, the basis varying 'according to the state of the times and . . . the price of provisions'.[60] By the 1870's the company's subsistence payments amounted to over 50s a month, compared with the Beaumonts' 40s (see Table 1). By far the greater part of the miners' earnings, therefore, came in the form of regular and guaranteed subsistence.

One final factor that should be mentioned was the increased productivity of the larger mines in the nineteenth century. Improvements in the overall design of mines and the mechanisation of ore dressing increased output considerably. It was this that enabled the owners to increase bargain prices without equivalent increases in the price of lead.

The movement of earnings in the period

Despite the fragmentary nature of the evidence, it has been comparatively easy to analyse the factors governing the earnings of lead miners. To give a meaningful chronological survey of their fortunes, however, is a different matter; gaps in the evidence, combined with the peculiar method of payment, do not permit this to be done in any great detail.

In the period 1720 (when the Weardale records start) to 1761, the normal top price in the Blackett concern was 25s. The average market price of lead fluctuated between £14 in 1730, £10 in the early 1740's and £12 in 1761. Unfortunately the letter books of the Blackett mines do not survive for the period 1742–48 when lead sold at the very low price of under £11 a ton. During more normal periods, for which letter books do survive, the miners complained of delays in payment, of both the general pay and the periodical subsistence. At the end of this period, when Jars paid his visit in 1765, the lead miners were paid on a scale at least comparable with the neighbouring colliers. Even allowing for gross exaggeration and the fact that he came in a period of prosperity, Jars's estimate of payment 'from 12 to 40 pounds sterling' in three months compares favourably with Adam Smith's estimate of the Tyne colliers receiving not more than tenpence or a shilling a day about 1763.[61]

After 1762, top bargain prices rose higher than ever before, but they fell again with the price of lead. By the 1780's subsistence was a fixed sum paid fairly regularly. The 1790's saw a great increase in bargain prices and in subsistence. The Beaumont chief agent wrote in 1795, when the miners of Allenheads asked that 5s a year should no longer be automatically deducted from their earnings for the Allendale clergyman, that 'the Wages of the Miners have for some years past, been advanced on a greater proportion than provisions have advanced in price and it has been found by experience that the more concession is made to them the more unreasonable they are'.[62] This 'unreasonableness' was prompted by the scarcities of the times, but also the lead miners were no longer paid on a scale comparable with the coal miners, and numbers of Weardale miners migrated to the north-eastern coalfield.

The period 1800 to 1810 was one of prosperity, with only minor slumps. Demand for lead because of the war sent its price up to unprecedented levels. All the mining companies raised their bargain prices accordingly.

In 1810–25 the lead industry went through a severe slump, followed by a partial recovery. Lead prices dropped towards the end of the war, and fell disastrously in 1814–17. The Beaumonts cut their top price down to 30s, the same as in 1770, and subsistence from 30s to 20s a month. Many men were forced to emigrate. Prices recovered before 1820, bargain prices were increased again—although not to their wartime level—and the 30s subsistence was restored. 1826 to 1836 was another period of severe depression and low prices. Earnings went down to little better than starvation level, and a substantial proportion of the population emigrated from the district.

From the late 1830's to the early 1870's the two major companies achieved a measure of stability in the face of fluctuating market prices. The London Lead Company gradually increased its monthly subsistence payments from 40s in the 1830's to 54s in 1872. The Beaumonts, who preferred to pay the larger part of their employees' earnings at the pays, increased subsistence to only 40s. Pays became half-yearly instead of annual in 1860. But wages were substantially below those given on the neighbouring coalfields.

In the late 1870's the market for English lead finally collapsed, with the importation of foreign lead at a price which the English mine owners could not match. These figures (based on export prices given in the *Statistical Abstracts of the United Kingdom*) show the fall in the price per ton:

1865	£21·05	1880	£17·41
1874	£22·63	1881	£15·72
1875	£23·17	1882	£15·45
1876	£22·55	1883	£14·07
1877	£21·49	1884	£12·58
1878	£18·74	1885	£12·25
1879	£15·42		

The mining companies were forced gradually to cut down the number of their employees, and the earnings of those that remained. The bitterness engendered by the 'new system' in the Beaumont mines has already been described. Both the big companies virtually suspended operations in the early 1880's.

The friendly societies

In times of bad health and in old age the lead miners had three possible sources of income. The first, parochial relief, will be dealt with in chapter IX. Second, both the London Lead Company and the Blackett/ Beaumonts gave pensions to retired agents, however inferior their status, and both also gave money to miners who had been badly injured while actually in the mines, or to their dependants if they were killed. Third, miners could insure themselves against illness or old age by becoming members of friendly or benefit societies.

Broadly speaking, there were three types of friendly society in the lead mining district: (1) small local societies, many probably being short-lived; (2) in the nineteenth century there were larger societies backed or organised by the mining companies; (3) again in the nine-teenth century, there were branches of national societies, such as the Independent Order of Rechabites and the Manchester Unity of Oddfellows.

The names and rules of most of the small independent societies have not survived. The earliest may have been the Society of Miners in Aldstone-Moor, whose rules, dated 1755, were printed by Raistrick.[63] Similar organisations were probably formed in the other mining areas, and by the end of the century there were certainly others. There are records of small payments to the Loyal Miners' Club at Allenheads in the Blackett/Beaumont records in the 1780's and 1790's. The St John's Chapel (Weardale) Friendly Society, founded in 1787, was still in existence in 1847, a much longer history than most of the small ones. [64] This was one of two friendly societies in Weardale in 1797,[65] and the number had risen to three by 1809.[66] Many were very small indeed and might have been missed by surveys such as those of Eden and in the various poor law reports. In 1852 a group of miners at Coalcleugh, with the encouragement of Thomas Sopwith, published a pamphlet entitled *Practical illustrations of the benefits to be derived from well-conducted friendly societies, with reference more particularly to the New Societies established in East and West Allendale, as compared with the old benefit clubs yet existing in the immediate district.*[67] The names of the old societies were 'Brown's Club', 'Errington's Club', 'Thornley Gate Club', and 'The Private Club'. The writers noted that 'the last named Society has obtained its title in a different manner from that of the others, being held at a private house . . . while the others expend a sixth part of their yearly contributions for intoxicating drinks in the public houses at which they are held, in lieu of rent'.

Little information has survived about the size and administration of these societies. Broadly, they aimed to give weekly sums to sick members, to pay pensions to those over a certain age, and to give a few pounds to the widow on the death of a member. The 1755 rules of the Society of Miners in Aldstone-Moor laid down that members should receive 4s a week the first ten weeks of sickness or retirement, and 2s 6d a week thereafter; widows were given £3 on the death of member husbands. Nearly a century later, in 1847, the St John's Chapel society had precisely the same benefits for its members—the 2s 6d a week pension was given at the age of 60. Conviviality, on the occasion of the periodic meetings of the clubs, was also a prime aim, as the rules of the Aldstone society show. Unfortunately, the number of miners who were members of such societies in the eighteenth century cannot even be guessed at.

The basic problem of the friendly society was the achievement of financial stability so that income and investments were enough to pay for benefits offered. Unless the society achieved this stability, accompanied by a constant stream of new members, it would fail, and members who had contributed subscriptions for most of their lives had nothing to receive when they hoped for benefits. The societies were constantly running into financial difficulty. The situation of the St John's Chapel society in 1847 was typical. Its income in the four years before had been £1,076; its expenditure on benefits £1,080. There were virtually no savings, so all income was immediately paid out as benefits. Young members were no longer joining. The society had only 53 members under 30 years of age, and 259 above. Such financial difficulties encouraged the more progressive mining companies to step in and arrange the benefit societies' affairs for them.

Both the London Lead Company and the Blackett/Beaumonts made occasional donations during the eighteenth century. In the early nineteenth century the Lead Company formed a society expressly for its workmen. In a resolution of 1810 the court of the company established a fund for the 'Relief of maimed and decayed Workmen employed by the Company'.[68] It was run on traditional benefit society lines by the men themselves, with an annual subsidy from the company. In 1827 the fund became insolvent, as 'the payments out were too large for the fund'.[69] An investigation by the court followed, and in October the balance of the fund was transferred to the general account of the mines.[70] From then on, membership was made compulsory for all employees, and the superintendent of the company and the district agents became *ex officio* members of its governing committee.[71] The company made a regular donation of £200 a year.[72] Robert Stagg, the superintendent, told the 1834 Poor Law Commission that 'though nominally maintained by the men, it is considered, and apparently with justice, to be in times such as the present really supported by the masters, since, having come to be considered as a necessary item of outlay, it is provided for in much the same manner as candles, gunpowder, or any other article indispensible for working.'[73]

The annual subscription of 30s for each man was deducted from his earnings at the pays. Benefits in 1834 were, to the sick, 8s a week for the first six weeks, and 6s thereafter; to members aged 65 or over, 5s a week.[74] In 1864 the benefits were the same, save that the initial sickness payments had gone down to 7s a week. Five pounds was paid to widows on the death of a member.[75]

The incidence of pneumoconiosis ensured that the pension was seldom claimed. In 1864, out of 1,083 members, only sixteen were over 65 and receiving a pension. In 1885, when the company was on the brink of dissolution, the fund had £58,170 14s 7d in investments.[76]

After Sopwith's appointment as chief agent, the Beaumonts too took a closer interest. Two new societies were formed—the Allenheads Benefit Society and the Allendale and Whitfield Neisonian Benefit Society—on the basis of rules drawn up by Sopwith. They were subsidised by the mining concern: 5 per cent of the total annual subscriptions and 2 per cent of the total capital invested were donated. Subscriptions were based on a sliding scale. Those joining at the age of 15 paid 9d a month: those at the age of 40, 1s 6½d.[77] Benefits were in line with those of the London Lead Company, but no retirement pension was given until the age of 70. In 1864 not a single one had ever been claimed!

Membership was not compulsory, and it would appear that only a minority of the Allendale miners were members. (Membership of the complementary medical scheme, providing free medical attention for sick workmen and their families, was compulsory, however.) In Weardale there was no similar society supported by the Beaumonts; the miners would not agree to the rules proposed by Sopwith.[78] There, and in the other mining districts, any friendly society that existed did so without the owners' support.

In 1864 there were no friendly societies at all for the Derwent miners, or at Stonecroft and Greyside, north of the Tyne.[79] In Weardale and Alston, however, there were still a number of small independent ones, and branches of the Rechabites, Oddfellows and Ancient Druids. In Allendale too there were societies other than the officially supported ones in 1864. Some miners were members of more than one society. Some of the Lead Company's miners were Oddfellows too, although this was discouraged by the company.[80]

In all, probably some two-thirds of the lead miners were members of friendly societies by the second half of the nineteenth century. Their chief value lay in the security they gave the sick and maimed. The retirement pensions were of little use; Robert Stagg himself, the probable originator of the comprehensive London Lead Company scheme, told the 1832 Poor Law Commissioners that 'the average duration of life' was 'just twenty years less' than 65. There was no attempt to give support to widows and fatherless children, of whom there were a large number, owing to the early mortality of the men. No money could be paid at times of unemployment. But even this limited amount of social security was better than the condition of the average industrial worker in the England of the time.

NOTES

[1] 1842 report (Mitchell), p. 744.
[2] B/B 52, 23 September 1833.
[3] B/B 47, 2 June 1756.
[4] B/B 50, 22 April 1793.
[5] B/B 51, 16 March 1812.
[6] 1842 report (Mitchell), p. 678.
[7] Sopwith, *Diary*, 24 June 1864.
[8] B/B B.M. letter book, 23 October 1730; 10 April 1731; 8 December 1732.
[9] B/B 46, 19 September 1758.
[10] B/B 46, 2 March 1759.
[11] B/B 46, 3 November 1759.
[12] B/B 46, 13 January 1761.
[13] L.L.C. 13a, 6 October 1785.
[14] B/B 52, 3 October 1833.
[15] B/B 46, 8 June 1759.

[16] B/B 48, 24 August 1766.

[17] B/B 50, 23 November 1795.

[18] B/B 51, 4 September 1816.

[19] B/B 51, 25 September 1818.

[20] L.L.C. 21, 17 March 1831.

[21] B/B 48, 24 August 1766.

[22] B/B 50, 17 December 1795.

[23] 1842 report (Mitchell), p. 766.

[24] 1864 report (vol. 2), p. 381.

[25] 1842 report (Mitchell), p. 744.

[26] B/B 53, 18 October 1811.

[27] B/B 96.

[28] 1864 report (appendix), p. 326.

[29] L.L.C. 32, 4 May and 4 June 1872.

[30] Sopwith, *Diary*, 15 September 1845.

[31] In his *Diary* for 29 June 1871 Sopwith noted that in the 22 years 1846–67, £16,780 of arrears had been cancelled. Only £260 of this had been met by the company: all the remainder had come from the 10 per cent levy on successful workmen.

[32] Quoted in J. Lee, *Weardale memories and traditions*, 1950, pp. 267–8.

[33] 1842 report (Mitchell), p. 744; 1864 report (appendix), p. 325.

[34] 1864 report (vol. 2), p. 346.

[35] B/B 49, 3 December 1777.

[36] J. Losh, *Diary*, vol. 2 (Surtees Society publications, vol. 174), 1959, pp. 58–9.

[37] Sopwith, *Diary*, 4 January 1847.

[38] Ibid, 12 November 1862.

[39] Ibid, 10 February 1858.

[40] Ibid, 15 November 1861.

[41] Ibid, 25 September 1845; 23 March 1847.

[42] J. W. Allan, *North country sketches*, 1881, p. 46.

[43] L.L.C. 13a, 16 October 1785.

[44] B/B 49, 19 March 1765.

[45] B/B 50, 8 November 1792.

[46] B/B 50, 9 May 1797.

[47] Featherston, 1840, pp. 60–1.

[48] Ibid, p. 745.

[49] The earlier figures are taken, by permission, from those quoted by Hughes. They give the average price of lead sold in every year. Tooke's figures are based on the prices obtained only in the last quarter of each year.

[50] Jars, vol. 2, pp. 544–5: '. . . though profitable to the interested parties it is, in the long run, to the disadvantage of the mines.' 'From these mistakes in fixing prices it follows that the earnings of the workmen are often much

more considerable than is proper: each of them earns in the course of three months from £12 to £40 sterling, which is too much. The proprietors say in answer to this that it is perfectly equal to them, because they have calculated so that the produce of the ore is always 50 per cent or greater, never less, and besides the system serves as an encouragement. This may be true in one sense but in general a workman who gains too much is seldom a good workman.'

[51] L.L.C. 16a, 31 October 1808.

[52] L.L.C. 16a, 24 June 1815.

[53] B/B 55, 8 July 1831.

[54] L.L.C. 21, 26 March 1829.

[55] L.L.C. 14, 7 November 1793.

[56] B/B 14, 51, 31 December 1814.

[57] L.L.C. 15, 20 June 1800.

[58] L.L.C. 15, 20 June 1800.

[59] L.L.C. 16a, Midsummer, 1817.

[60] 1864 report (vol. 2), p. 375. For an account of the different basic wages, see A. Raistrick and B. Jennings, *History of lead mining*, 1965, pp. 293–6.

[61] T. S. Ashton and J. Sykes, *The coal industry of the eighteenth century*, 1929.

[62] B/B 50, 28 January 1795.

[63] Raistrick, 1938, pp. 42–7.

[64] T. Sopwith, *Substance of an address to the members of the St John's Chapel Friendly Society*, 1847.

[65] Eden, vol. 2, p. 168.

[66] J. Bailey, *Agriculture of Durham*, 1813, p. 327.

[67] A possibly unique copy of this tract is in the Cowen collection, Newcastle University library.

[68] L.L.C. 17, 6 December 1810.

[69] 1864 report (vol. 2), p. 378.

[70] L.L.C. 20, 1 November 1827.

[71] 1842 report (Mitchell), p. 749.

[72] 1864 report (vol. 2), p. 378.

[73] Poor Law Commission report, 1834 (appendix A), p. 139A.

[74] Raistrick, 1938, p. 52.

[75] 1864 report (vol. 2), pp. 377–8.

[76] L.L.C. 34, 17 September 1885.

[77] *Practical illustrations of the benefits . . . from . . . friendly societies*, 1852.

[78] 1864 report (vol. 2), p. 371.

[79] Ibid, pp. 346, 355.

[80] Ibid, pp. 331, 378.

V

Washers, smelters, carriers and agents

After extraction, lead ore has to go through a number of processes before it becomes lead metal. All impurities have to be removed by dressing, or 'washing' as it was universally called in the northern Pennines. Roasting and smelting follow, to transform the lead sulphide content into lead. By the eighteenth century, specialised groups of workmen were carrying out the latter process, and by the third decade of the nineteenth century washing too had become a specialised occupation.

Washing operations were carried out as close to the mines as possible, to avoid the unnecessary transportation of worthless impurities; but most smelt mills were further away, on the edge of the mining area, so as to be near an adequate supply of fuel. The ore was carried from the washing floors to the smelting mills by carriers, who used trains of packhorses (called galloways) over the roadless fells, and small carts on the new roads built from the end of the eighteenth century onwards. Before the coming of railways, they were also responsible for getting the lead out of the mining area to the marketing centres of Newcastle and Stockton. Both the actual extraction of the ore and all these subsequent operations were supervised by the officers of the various concerns interested in lead mining—the agents.

Washing the ore

By the mid-nineteenth century, a vast increase in efficiency had made many mines and veins previously considered too poor to bother about worth working, and mechanisation had completely altered the organisation of the washing operations. Washing the ore became almost a technology in itself, with its own specialised vocabulary. Among the machines and implements used were 'dolly tubs', 'buddles', 'buckers', 'colrakes' and 'limps', whilst ore at different stages in the operations was known as 'chats', 'cuttings', 'smiddum', 'smiddum tails' and 'slimes'.

It is necessary to give a brief account of this technological revolution in order to understand the consequent revolution in organisation which changed the labour force from many small groups of washers working independently to a single large group working as a team at each mine under factory-style regulation.

The object of washing the ore was to separate from it the useless stone and spar with which the galena, or lead sulphide, was intermingled. To a certain extent this could be done by eye alone, but much galena would be lost if no further effort were made. The methods of separation were based on the fact that the specific gravity of the galena was greater than that of the impurities mingled with it. If ore were dropped into a column of water, it should separate into three layers, the galena sinking fastest and forming the bottom layer; then would follow a mixture of galena and stone, and lastly the stone would lodge on top. For this to work in practice, it was necessary that all the pieces of 'bouse', as unwashed ore was called (when washed, it was called 'ore'), should be of the same size, and it was also found that shaking the bouse in water induced separation. The first operation, therefore, was to smash or crush the bouse into pieces of a uniform size; the next, to carry out the separation in water. The revolution in washing practices at the beginning of the century consisted of a vastly more efficient crushing process that produced a greater uniformity in the size of the particles of bouse, and of a more sophisticated (as well as mechanically driven) separating apparatus that allowed the steadily decreasing mixtures of galena and impurities to be washed and re-washed again and again.

For most of the eighteenth century, washing practices in the area were extremely backward, compared not only with those employed on the Continent, but also with those in use elsewhere in Britain. Jars, inspecting mines all over Europe on behalf of the French Government in 1765, visited the lead mines at Leadhills, in Scotland, and on Alston Moor. At Leadhills he discovered that the washing was carried out with insufficient care, and the techniques used were very inefficient compared with German practice. Much ore was unnecessarily lost or thrown away. In the northern Pennines it was the same as at Leadhills, 'mais encore avec beaucoup moins de précaution; de sorte que l'on perd quantité de minérai, indépendamment de celui que l'on jette au rebut.'[1] The stamping mill for crushing very hard ore, known to Agricola in the sixteenth century and in use at Leadhills, was not employed at the washing floors of Alston Moor or Coalcleugh, although it was used in the smelting mills for crushing slag from about 1750 onwards. Jars noted, however—in fairness to the miners of the northern Pennines—that the system of leases, whereby a certain percentage of the ore that was raised went to the landlord, made it uneconomic for the mining companies to wash bouse again and again for the sake of the ever-decreasing and qualitatively inferior ore remaining.

At this time the method of washing was as follows. The lumps of bouse were first broken up into fairly uniform fragments by hand beating with a 'bucker', a mallet-like instrument consisting of an iron head on a wooden shaft. The purer fragments of ore were removed, ready to go to the smelt mill. From some 'earthy' veins perhaps 60 per cent or more of the ore could be separated ready for smelting after this one operation. In other veins, where the galena was much mixed with stone and metallic impurities, very little was ready after this stage.

The crushed ore was then placed in a hand sieve and shaken in a tub of water, or a suitable pool in a stream. The bouse gradually stratified into three layers—pure ore at the bottom, a mixture of ore and stone known as 'chats' in the middle, stone on the top. The top layer of stone was swept off with the piece of iron known as a 'limp', the ore was put aside, and the chats were again crushed and sieved. The process was continued until all the ore had been extracted, or the patience of the operator exhausted—the latter, according to Jars, occurring very early by Continental standards. The 'smiddum' (fine particles that fell through the sieve to the bottom of the tub) was then washed on a 'buddle', which in its simplest form was a few planks of wood laid at a slight angle on the hillside. Water was run down the buddle in a continuous stream; the fragments of smiddum were dropped into the flow and stirred with a rake. This process again separated it into three layers with pure ore at the top of the buddle, while the mixed stuff and waste was swept further down, or washed off altogether. The pure ore was removed and the 'middles' washed again, as often as necessary. More complex forms of buddle had been developed by the end of the eighteenth century but the principle remained the same. Washing equipment in use about 1800 is shown in plate IV.

This was obviously a crude and imperfect method of dressing the ore. Much was lost, and sufficient remained in the heaps of deads and waste material for partnerships to be constantly kept busy re-washing them. The crude method of hushing was used, the violent flow of water separating ore and waste. Jars wrote that ore was found in stream beds miles below the actual mines and washing floors, and that bargains were taken for gathering it up. In a letter concerning the proposed Weardale enclosures in 1799, the Beaumont chief agent wrote that should it take place 'we should not be deprived the liberty of hushing Wastes as it is not only the best way of working them, but also does least damage to the Ground, as by that means they are continually conveyed to some Burn or Rivulet, and by any other method of washing them they are spread, and left upon the Ground'.[2]

In 1796 'an improved method of washing ores was introduced by Richard Trathan, a Cornish miner who came in search of employment, and proposed to obtain considerable quantities of ore from the refuse of the then inexperienced washers'.[3] Working for the London Lead Company, he introduced two fundamental innovations, long known elsewhere, into the area. First, he used a stamping mill to crush ore too hard to be broken by the buckers. Second, he dug what were known as 'slime pits' below the buddles to catch the fine particles of ore that had been washed right off the buddles and, previously, lost.[4] Sopwith noted that Trathan's bargains 'produced him such good wages that the agent endeavoured to reduce them'. Upon his resisting, Trathan was dismissed 'and a whole summer passed in the fruitless efforts of others to supply his place'. He was then re-engaged. Trathan's work was the foundation of such an improvement that other French inspectors visiting the area about 1830 thought its washing methods were as advanced as those used in Germany.[5]

By 1842 the process of washing at mines belonging to the larger concerns had been mechanised. Compared with eighteenth-century practice, it was highly efficient in obtaining every scrap of smeltable ore. At the smaller mines, where the owners or adventurers had no capital to invest in machinery, things went on much as before. The efficiency of the new system lay in the mechanical crushing apparatus, which reduced all the fragments of bouse to a uniform size, and in the number of processes the chats were subjected to. At every succeeding stage, the ore obtained was of slightly inferior quality, but improvements in smelting techniques meant that it could still be dealt with profitably.

To contemporaries it was the first operation, the crushing, that brought about the most striking improvements in productivity. Without this the more sophisticated machinery for separating the ore and the waste would have been much less effective. The 'crushing machine,' two or three fluted rollers driven by water power, revolving against one another and crushing ore fed in from above, was first introduced from Cornwall by Ulrich Walton & Co. at Alston Moor in 1797.[6] The London Lead Company and the Beaumonts soon followed suit. The Lead Company's chief agent calculated in 1811 that the machine 'will make at least $\frac{1}{5}$ more of the ore than the common way of washing'.[7] It was both more efficient than the buckers and required far fewer men to work it. By 1830 there were machines at all the mines of the two large concerns.

The separating methods were also partially mechanised, although a large labour force was still required, as the actual sorting of the ore after each stage of treatment had still to be done by hand and eye. The use of a water-washed grating for this purpose lessened the labour required. The sieving, or 'hutching', was now done easily by a single boy, as the sieve was suspended above the tub and moved by a long pole which also acted as a counterbalance.

Buddles became more sophisticated. By 1842 many different forms were in use, some power-driven, all designed to separate the ore and wastes. The cloth separator came into use in the late 1840's. It consisted of a continuous belt which both moved along and shook from side to side, thus throwing off the wastes and carrying the fine smidden with it. The washing floors at large mines were systematically planned to allow the ore to pass easily from operation to operation, with the railroad from the horse level bringing the bouse to the starting point. Dams were built to provide an adequate supply of water even in times of drought.

This mechanisation of the washing had three main effects. It made existing mines more profitable, since less galena was lost. This happy result, however, was chiefly to the advantage of the large companies. Small groups of adventurers could not afford the capital outlay and continued to work in the traditional manner. Second, mines or parts of mines previously too poor to work because of the low quality of the bouse became worthwhile. This was to the good of the men as well as the masters, since it increased the employment available. The Weardale agent for the Beaumonts reported in 1828, after the erection of a crushing mill at Kilhope mine, that 'The Workings are poor but the produce will be increased from that of last year in consequence of more Men raising Ore, a part of which have taken workings for the purpose of drawing old deads and washing them, which the use of the Crushing Mill has enabled them to do, and previous to the Mill being built, such places would not Work at the highest prices given.'[8]

The third effect was completely to change the organisation of washing. With the introduction of machinery it became necessary to have a specialised labour force, instead of each partnership being responsible for washing its own ore.

The earnings of each partnership, it will be recalled, were based on the total number of bings of *dressed* ore raised. This system of payment persisted even after the partnerships raising the ore ceased to wash it themselves.

Because washing was the responsibility of the bargain takers, there is little documentary evidence of practices in the eighteenth century. The masters were concerned only with the quality of the dressed ore, not with how the work was done. The washing agent watched to make sure that all impurities were removed, and that there was no cheating when it was finally weighed out in bings. His status was low; the Beaumont chief agent wrote of an agent in 1797 that 'he is a poor ignorant fellow and is only fit to look after the Washers and to deliver out the Gunpowder and Candles.'[9] Whenever the washing was unsatisfactorily carried out, it was the agent who was blamed; 'as there has been repeated complaints of the Lead Ore being not well dressed . . . I took the opportunity to go round the different Mines with the Washing Agents and examined each parcel, and was sorry to find some of them so dirty that I was under the necessity of informing the Washing Agent that unless these parcels were again washed over (which they promised to see done) we could not receive them at the Mill.'[10]

For most of the eighteenth century, washing was carried out either by the partnerships themselves or by their directly employing others to do it. The distinction between the two systems is not as obvious as would appear at first sight. It was often the sons or wives of the partnership members who did the washing. Jars said that in 1765 the partnerships themselves performed the operation.[11] A number of bargains in the Blackett books instructed the workmen 'not to wash their last Quarter's Work but the Agents let it to wash to whom they appoint'.[12] At Fallowfield, where the payment record for 1769–70 shows how the partnerships spent their time during a given quarter, most men worked a number of days at the 'Washg. Place at Middle Whimsey'.

This haphazard system went on into the nineteenth century. A London Lead Company note of 1812, recording the work of a particularly productive partnership, stated that 'in the summer season they employ not less than 10 men and lads to wash the ore'.[13] A year before, the company's chief agent had reported that before the end of each bargain period 'The Miners come out of the Mines to assist the Washers in making up all the Ore against that day'.[14] Thus at the beginning of the nineteenth century the London Lead Company partnerships were still directly responsible for either doing the washing themselves or employing someone else. According to Forster, writing in 1821, the miners were charged by the washers a price per bing of washed ore which fluctuated according to the difficulties involved: rich bouse requiring little washing was charged at a low rate; poor bouse, with many impurities in it, was charged at a high one.[15]

The equipment was very simple in the eighteenth century. The miners were expected to find it themselves, as they were the tools they used inside the mine. One Blackett bargain for washing wastes in 1766 stated that the men were 'to have four new Buddles allowed them but to find the rest of their Work Gear themselves'.[16] Mechanisation, of course, changed all that. The operation had to be comprehensively organised to take full advantage of the potentialities of new machines. It was done in two ways. The mining company itself could take over all washing operations, charging the partnerships accordingly; or they could be allocated to one contractor in the same way as drawing work.

The London Lead Company took the first course about 1816. The change is recorded by an adult washer in 1842, remembering conditions at the beginning of the century: 'After some years [working as a miner] I left my brother and took to washing, and hired boys under me, and was paid by the bing. The people who employ boys get more money than the men. The Company after some years put us all upon wages, and employed the boys themselves.'[17] The partnerships were charged in the same way as before, according to the quality of the bouse and the amount of work needed. The charges were deducted from the sums due to each man at the pays.

Other mining companies—most, according to Dr Mitchell in 1842—employed the 'contracting system':

A contractor at the washing-floor of the Derwent Company was pointed out to me, and I entered into conversation with him. He told me that he had 40 persons working for him. I said to him, 'At that rate you ought to be making a little fortune?' He replied 'No, no, Sir, I am watched too closely for that; if they see that I am making more than they think fit that I should do, then when my contract for this quarter is out they will reduce me the next quarter'.[18]

The contractor had to disclose to the mining company all prices charged, and all work done; his employees were paid at rates laid down by the company. 'After all it is a complete disclosure of the amount of his profits, and he is not very likely to make beyond the wages of an overseer under engagements so made.'

Appendix 5 is an example of a Beaumont washing contract. These contracts first feature in the Beaumont records in 1833, at a time of extensive administrative change within the concern. The contract laid down the wages to be paid to the washers, the hours of work, some instructions as to methods to follow, and a clause prohibiting the miners 'to render the Washers any assistance in the Washing of their own Bouse'. The clauses governing the contract were so comprehensive as to make the washers virtually the direct employees of the mine owner.

One indirect effect of the mechanisation of washing concerned the bargain system, and has already been touched upon in chapter III. The traditional system of payment by bing became very 'inconvenient for dressing operations, as after each parcel produced by the several partnerships . . . is washed up, the different apparatus must be stopped and emptied.'[19] In 1870 the cost per ton of washed ore was 7s 6d if this had to be done; when the ore could be treated as a whole, the cost went down to 2s 6d. Thus an increasing number of the larger concerns went over to the fathom method of payment, which did not require every partnership's work to be washed separately.

The Blackett/Beaumont records show that in both the eighteenth and nineteenth centuries, certain partnerships were specialist 'waste washers', working outside the normal washing system. These were partnerships taking bargains to 'wash up' the cuttings, slimes, deads, wastes, etc. They were on a tontale basis, presumably because the ore produced was of a poorer quality than that normally raised direct from the mines, and a given quantity would produce less lead. These contracts are recorded from the early eighteenth century right through to 1880. The number of partnerships engaged in this work was far fewer than those doing the actual mining, averaging perhaps three or four per cent of the total.

The actual work of the washers remained in essentials the same before and after the introduction of machinery—i.e. it consisted chiefly of sorting wet lumps of stone and ore. Before the crushing machine was introduced, a larger proportion of the labour force was engaged in breaking large lumps manually with the bucker. A 41 year old washer remembered in 1842 that when he first went to work at the age of 11, 'At that time crushing-mills were not so much in use as now, and boys broke the stone to separate the ore with buckers . . . ; buckers are now not much use except in small concerns.'[20] Other witnesses told the commission of their daily routines. 'I picked grating, that is, I took away the ore from the stones lying in the grate . . . I next went to hutch, that is, I took the stuff off the sieve after the ore was gone out of it.'[21] 'When I first went to washing I turned the handle of the thing in which they wash the stuff, the buddle. I next went to drawing slime. There is a trunk and water comes over it, and a boy puts in the slime, and we rub it with a colrake, and the water runs through it, and washes away all the mud, and leaves the lead.'[22]

The work was not hazardous—serious accidents were practically unknown—but it could be very uncomfortable. Dr Mitchell commented, 'the work is not too laborious'. The younger boys were given jobs within their strength. The worst feature was the lack of protection from the weather.[23] On the exposed Pennine fells, working in the open with no form of shelter could be most unpleasant, even in the summer months. Work went on through the rain, and even if the rain stopped there was nowhere for the workers to dry out. Arms and hands constantly wet were made raw by the wind. The only place where Dr Mitchell saw sheds over the washing floors was at Coalcleugh; there were none at any other Beaumont mine. The London Lead Company occasionally provided boards to act as wind-breaks, but never any overhead covering. The commissioners strongly recommended that such covering should be provided. By 1864, however, little had been done except by Sopwith, who had built sheds at most of the Beaumont mines. The Lead Company had shelters at some, but not at all mines.[24] The remainder had nothing at all.

Because of the relatively light nature of the work involved, washing was traditionally not a job for grown men, except in supervisory capacities. If the partnership delegated the work, the washers were frequently the sons and wives of its members. Many women were included in the lists of waste washers taking bargains in the Blackett records of the eighteenth century. Some partnerships were composed entirely of women, others were husband-and-wife or father-and-daughter partnerships. By the end of the century, however, women's names became increasingly rare, and no woman is *listed* as having taken a washing bargain in the nineteenth. In 1817, a time of great food shortage, women besieged the London Lead Company agent, asking for work;[25] this may mean that they had been regularly employed until then, or that the famine had driven them to seek work. In 1834 no women were employed in washing in any of the lead mining parishes which furnished returns to the Poor Law Commission.[26] In 1842 the commissioners found only two employed at washing in the whole area, and miners, questioned as to the reason, considered the idea 'very improper'.[27] In 1842, therefore, it must have been many years since women had been generally used. The reason is not obvious. Early nineteenth-century morality was not normally shocked by the idea of women working alongside men, at least not to the extent of putting a stop to it. There was no alternative work in the area, and at the Yorkshire lead mines, as well as in Cornwall, women were still dressing ore in the mid-nineteenth century.

The bulk of the labour force consisted of boys under the age of 19, since the future miners normally began their working careers as washers. In 1842 most boys started work between the ages of 9 and 12. 'Children under 9 are seldom so strong as to be of any use whatever, which is the best security against their being employed.' The London Lead Company usually took on no one before his twelfth year, but 'the importunity and poverty of the parents, particularly of widows, procures the relaxation of the strict rule.' By 1864 most companies refused to accept children under 11. Bainbridge gave the 1864 commission an account of the washer's normal career and his progress towards becoming a miner:

From the age of 12 to 18, as a rule, they must attend the ore-dressing floors, and at 18 years of age they are allowed to be placed underground. At 14 years of age we allow them to go underground during the three winter months when the ore dressing is not in operation . . . . He is placed with one of the mining partnerships as a labourer, and in that way he is serving a sort of apprenticeship, or obtaining knowledge that is likely to be useful to him in after life. At 18 years of age we consider these young men eligible to be placed as partners in the mines, and to become members of a partnership. . . . In recent years, owing to the fact that we have had a greater amount of ore to dress than we had boys under 18 to dress it, they were kept out longer, so that the bulk of our boys may be said to be kept out of permanent underground working until they are 20 years of age, and even beyond that.[28]

The severity of the Pennine winter meant that washing normally ceased between November or December and April; even in summer the weather was often bad enough to bring work to a halt. Before the construction of dams, drought had the same effect. 'Altogether a washer is prevented from working above 21 or 22 days in the month, and he works at washing from eight to nine months in the year. In the wintertime, when the washing becomes impossible, many of the young persons go to work in the mines and the young boys go to school.'[29]

Washing therefore had to be crammed into rather less than eight months of the year, while mining went on all the year round. This was partly why the hours of work were so long. It was a feature of contemporary coal mining practice, too, that the children worked many more hours than the adults. In the Beaumont washing contract of 1833 (see Appendix 5) the normal working day was 7 a.m.–7 p.m. with an hour for lunch. On Saturdays (when the adult miners did not normally work) work ceased at 12.00, but if any time had been lost through bad weather during the week, it could go on during the afternoon as well. In 1842 similar conditions applied at all the mines. Towards the end of the washing year, about September, work continued until midnight on fine days. At Allenheads 'it falls on each boy once or twice-a-week, excepting some little boys, to stay till midnight, from 7 A.M.'[30] In 1864 hours of work were still much the same.

The miners were charged by the piece, according to the work involved in preparing each bing. In 1821 the charge could be as little as 2s 6d for exceptionally clean bouse, or as much as 8s for a very hard mixture.[15] When the mining companies took over the washing they kept the old system (unless fathom working was also introduced) but paid the workers by the day. 'When children go to wash at nine years of age the usual wages are 4d. or 5d. a-day, and it is customary to advance the child a penny a day for every additional year of his life. Some years there is a rise of 1½ or 2d. a-day.'[31]

The statistics of workers employed in the mines on Alston Moor, supplied by the different companies to the Greenwich Hospital (see Appendix 6), show that the proportion of washers to miners went up in the nineteenth century. Between 1738 and 1767 the number of washers in the total labour force (taking the quarter of the year when the greatest number of workers were active) varied from under 10 per cent to as many as 20 per cent. In the years 1818 to 1844 the washers almost invariably formed 30–36 per cent of the total. This proportionate increase, in spite of the partial mechanisation of washing, shows that in the eighteenth century many miners must have washed their ore themselves.

Smelting

Once the ore had been dressed, it was ready for smelting. The basic principle was to reduce the ore—mainly galena, or lead sulphide—into

pure lead by heating it in a blast of air. The oxygen in the air oxidised
the sulphur, leaving the lead behind. No more than a bare outline of the
processes is called for here because, unlike the process of washing, no
technological change occurred in smelting to alter the structure and
administration of the industry. The smelters were specialist, highly
skilled workmen in the eighteenth as well as in the nineteenth century.

The smelting mills were frequently separated from the mines
geographically as well as administratively. The siting of smelt mills was
dictated by three factors: smelted lead was roughly two-thirds the
weight of the ore, and was thus cheaper to move; fuel was not easy to
come by on the high moorland where most of the mines were situated;
ease of access to the marketing centre—Newcastle or Stockton—had to
be considered. The smaller mining organisations tended to locate their
mills as near their mines as possible, to avoid extra transport costs, and
to use only peat as fuel. The larger ones either undertook the cost of
taking wood and coal to their mills—as the London Lead Company
did at Nenthead—or compromised by siting them where wood and
coal were available, but not too far away from the mines—as the
Blackett/Beaumonts did at Dukesfield and Allen (Allendale Town)
mills and the Greenwich Hospital at Langley. The Blackett/Beaumonts
refined their lead at Blaydon, near Newcastle. The other concerns
attached their refineries to the smelt mills. Most of the bigger companies had more than one mill, to cover activities in different districts.[32]
The smelters themselves tended to live apart from the miners. At
Nenthead they occupied a distinct section of the village. The 1851
census shows that smelters' sons became smelters also, rather than
transferring to the extractive side of the industry. In their boyhood
they worked as washers alongside the miners' sons but as soon as they
were old enough they became smelters.

All the smelting operations were carried out in furnaces in specially
built smelt mills, sited near water, which supplied the power for blowing
the bellows.[33] The most common mode of working throughout the
period was to smelt the ore in an ore hearth—a furnace where the ore
and fuel were mixed together rather than kept in separate chambers.
The ore was first 'roasted' to expel some of the sulphur. In the
eighteenth century this was done by putting raw ore on top of that
already in the hearth, and allowing it to roast before mixing it in with
the rest. In the nineteenth century it was usual, in the larger mills, to
roast the ore in a reverberatory furnace, where the fire was kept
separate from the ore.

The smelting proper was then carried out. Ore and fuel (usually peat) were mixed in the furnace and set alight. Air, blasted in by bellows was distributed throughout the burning mass by placing peat before the nozzle of the bellows. The whole mass soon acquired a pasty consistency, so much depended on the skill of the smelter in stirring up the 'browse' and bringing it forward upon a large slab in front of the furnace, known as the 'workstone', where large lumps would be broken up, and expended ore, now called 'grey slag', removed. The liquid lead flowed from the furnace by a gutter across the workstone into a pot, from which it was ladled to form pigs. More fuel and fresh ore would be added as necessary, and the smelting would go on for twelve or fifteen hours, after which the furnace became too hot, and had to be left to cool for a while.

Ore smelting was universal in the northern Pennines, and except in the mills of the London Lead Company it continued in use with minor modifications until the end of the nineteenth century. The London company adopted the reverberatory furnace for smelting proper as well as for roasting. This was more efficient in that the 'black' slag remaining contained no combined lead, as did the 'grey' slag from the ore hearth, and so did not need a second smelting. But reverberatory furnaces required coal to produce an intense enough heat; ore hearths worked best on peat, supplies of which were readily available.

The ore hearths would also deal economically with small packets of ore and would accept lower grades than the reverberatory furnace. But many grey slags remained after each shift, to be treated in a slag hearth, which produced a more intense heat. The fuel (coke) and slags were mixed and lit, and the whole mass became liquid as the heat increased. A blast of air was blown through by a bellows as before. When hot, the mixture was tapped into an ash-filled pit. The lead ran through the ash into the pit below, and the black slag caked on top of the ash. When solid, it was removed and broken with a stamp mill to extract any particles of lead trapped in it. The pieces were washed, and the remaining lead re-smelted.

Some ores were rich in silver as well as lead. The product of different veins was kept rigidly separate and, as the silver-bearing veins were known, only the lead from these was refined to extract the silver. There was a great technological advance in refining in the 1830's, when Pattinson's process was invented. Until then refining had been done by 'cupellation' in a reverberatory furnace. Liquid lead was heated in a container of bone ash until it was red hot; blowing air across the surface with bellows caused 'litharge' to form. The litharge was blown out of the furnace, and fresh lead allowed to melt into the 'test-bottom'. Four fothers, each of 21 cwt, were worked in this way. At the end there remained in the furnace 'rich lead' containing a high proportion of silver.

The process was repeated with more lead until a large quantity of rich lead had been produced. The rich lead was treated in the same way, leaving, eventually, pure silver. The litharge then had to be transformed back into lead by heating it with coal in a reverberatory furnace. The 'test-bottoms' of bone ash were also re-smelted. Pattinson's process avoided the second operation. The lead was melted and allowed to cool slowly; it crystallised before the silver, and was removed with a perforated ladle. Rich lead was left. When there was sufficient rich lead the process was repeated until pure silver was obtained.

An idea of the size and complement of a smelt mill is given by the statistics of the Greenwich Hospital's Langley mill in 1822; it was probably the largest in the northern Pennines at the time:

	No. of men
Three roasting furnaces: three men each	9
Seven ore hearths: four men each	28
Two refining hearths	5
One reducing furnace (to convert litharge to lead)	3
Two labourers for weighing ore	2
One smith, one wright, one driver (of three carts)	3
	50

In addition there were six men dealing with zinc from one mine on Alston Moor.[34]

The smelters and refiners needed a high degree of skill to carry out every operation successfully. There were no mechanical aids, no dials to indicate temperature or fluidity; all had to be estimated by the smelter on the basis of his skill and experience. In ore hearth smelting, for example, great care had to be taken that the blast was neither too weak nor too powerful, and 'the same blast is not suitable for every variety of ore'.[35] The blast had to pass equally through every part of the mixture of ore and fuel, and the ratio of fuel to ore, which varied with the quality of the ore, had to be correct. If the process went on too long, or not long enough, a great deal of lead was lost.

Refining was even more complex. Over all the operations the greatest care had to be maintained. In 1730 it was discovered after some pigs had been sold that someone at the Blacketts' Dukesfield mill had put 'lumps of slags into the pig pan, which will certainly be attended with the ill consequence of depressing our Lead and giving it an ill character at all marketts'. The chief agent wrote feelingly of the culprit that 'hanging is too good for him.'[36] Short of assaying every pig of lead, there was no way of checking on the skill and honesty of the smelters. The Greenwich Hospital receiver wrote in 1769 that 'giving an entire Confidence to the Workmen . . . is perhaps the only Method to keep them honest, as . . . should they incline to be dishonest . . . there is no preventing their taking advantages but by making Assays.'[37]

The Blackett/Beaumonts, who operated several mills, were constantly checking smelters' ability by smelting ore from the same vein at different mills. In 1809 smelters at Rookhope mill 'left off their work alleging as the reason that they could not make sufficient Wages to keep their families from the Ore being so bad.' The chief agent wrote 'that I was not sorry the Smelters had left their work as I believed they had some very bad ones amongst them.' Similar ore was sent to Dukesfield mill to see whether 'it was the quality of the ore or want of skill in the Rookhope Smelters that caused the deficiency of their produce and Wages.' It was found that although the Dukesfield smelters did obtain slightly better results, 'there certainly was just cause of complaint in the quality of the Ore.'[10] The men were given an increase in wages.

A good smelter, therefore, was worth his weight in gold—or at any rate in silver. A primitive system of industrial training was in use at all the mills. Boys were not employed. A young man started off as a labourer, and then became the second man in one of the two-man teams working at the furnace. If he were sufficiently good he would eventually become a refiner, the 'highest branch' of the profession.[38] Both the Greenwich Hospital and the London Lead Company encouraged their chief smelters to tour other mills, and the latter sent some employees on courses in chemistry at Durham University in the mid-nineteenth century.[39]

There was competition between the mills of the different mining interests to get the best men, and many smelters changed from one mill to another—something that happened less often on the extractive side. The inventor of the Pattinson process, Hugh Lee Pattinson, worked for the Greenwich Hospital, the Beaumonts and the London Lead Company during his career. The Greenwich Hospital correspondence at the time of the setting up of Langley smelt mill in and after 1768 shows the difficulty of getting and keeping together a skilled labour force. Workmen had to be attracted from other concerns, and those already secured kept happy to prevent their seduction by attractive offers from elsewhere. The best period for piracy was also the danger period when the Hospital's own workmen might leave; this was the winter time, when no fresh ore was coming from the washing floors. In December 1770, Mulcaster, the Langley Mill agent, was told to allow the men to work as hard as they wished, and earn as much as possible, as 'their being allowed now and at all times to make good Wages will induce them to keep themselves disengaged til new Ore does come in'.[40] A year later, the receivers wrote to Mulcaster 'that the persons who were employed at Fallowfield in running their slags were at or about a finish there. At the same time it was intimated to us that it was likely they might be got to Langley Mill if we had occasion'. This would not only increase the mill's labour force, but there was also 'a prospect of our learning something from them'.[41]

Attracting skilled workmen from other concerns had to be done in an underhand manner. The practical necessity of mobility of labour was recognised, but openly enticing workmen by offers of better wages and conditions was still regarded as immoral. In 1771 the receivers wrote to Mulcaster:

The Work now most assuredly requires rather a Workman of experience than a learner and therefore the necessity of the thing must occasion our getting such a Person; indeed we see upon all things of this kind that other People are not so scrupulous as we are and if we don't take such steps as to be enabled to have proper Workmen we shall none of us have any credit in the Undertaking. I would however have us to be careful, but not so scrupulous to others as to neglect ourselves and upon the whole I look upon it under the present circumstances to be justifiable to take such person as thinks himself at liberty *to offer* himself to us and consequently it must be understood, that you should contrive to have it known in the Country, that a man is wanted but this with as little bustle as possible.[42]

Open approaches were not altogether unknown—in 1780 the Blacketts were asked if they would be prepared to exchange men skilled in one branch of smelting for some skilled in another.[43] The competition for skilled labour probably continued in the nineteenth century, but documentation is lacking.

Wages were the principal method of attracting and retaining smelters. There were other benefits; the smelters occupied the best houses in the London Lead Company's village of Nenthead, and it was found necessary to allow the Langley smelters land for smallholdings, with the refiners getting the best (see chapter VII), but wages were the most important factor. One of the Greenwich Hospital receivers wrote in 1771, 'I am not pleased to hear the Company intend raising their Smelters' Wages . . . However if they do advance I see we must follow, or they may perhaps get our best workmen from us.'[44] Comparative details of smelters' wages are unfortunately not obtainable, as no wage books survive, only a few rates for jobs not comparable with those of other mines at different times. The existing evidence shows that although all the mining companies paid their smelters on the basis of the weight of lead or number of pigs produced, there were many marginal differences that could have attracted or repelled workmen.

The London Lead Company and the Greenwich Hospital paid their smelters by the fodder (21 cwt) of lead produced; the Blackett/ Beaumonts paid by the ton. This difference alone makes comparisons difficult, but there were many others. The most important was the variable quality of the ores worked. There was a clash of interests here. The miners naturally wanted their ore to weigh as much as possible at the end of the washing operations—from their point of view, it should be washed just sufficiently to satisfy the washing agent and no more. The smelters, on the other hand, wanted it as pure as possible; impurities affected the quality of the final product and slowed down all their operations. Some grades of ore, and ore from certain veins, always tended to be more impure than others. In 1800 the Greenwich Hospital was giving its men an allowance varying with the quality of the ore; the London Lead Company did not do this.[45] Another factor was the increasing efficiency of smelting processes, which enabled more lead to be smelted in a given time. Prices were reduced on this score. Recessions in the lead trade also caused temporary cuts in the price per fother or ton, just as bing prices for the ore raised were cut. Yet another factor was the number of men working at a hearth. Referring to the men's complaints of low wages at the time of the 1809 strike at Rookhope, the agent wrote:

I find they have earned upon an average this year about 10/9 each per Week, but being now over many Men to each Hearth is part of the cause, the Wages divided among the proper number would be about 14/6 for each Man. There is now the proper complement of 4 men to each Hearth. The smelters' earnings at Dukesfield have been about 9/8 each per Week, but there they have also over many men. . . . At Allen Mill the Smelters having constant work, and not over many Men earn about 14/6 per Week.[10]

In the Greenwich Hospital correspondence there are many references to the rates and wages of smelters that give a (very incomplete) account of changes and disputes from 1768 to 1830 at Langley mill. In 1768, at the time of its establishment, the smelters at the ore hearths were to receive 'about 6/– per fodder', giving them 'about 9/– per week wages'.[46] In 1772 the standard rate was still 6s, but after some complaints from the men that the smelters at Nenthead were better paid, an allowance was given for 'extra Labour' on poor quality ores.[47] The men were not satisfied with this arrangement alone, and the mill agent was urged to impress on them that 'tho' the Smelting Wages are the same now as twenty years ago, yet so much improvement has been made in the Smelting Art and Machinery for that purpose that the dispatch of business will in some measure countervail the increased price of Provisions'.[48] The smelters were not convinced. In September 1773 'we had a formal application from the Smelters in a body for an increase of Wages. . . . We found an increase necessary in order that we might keep our best Workmen who tho' they discreetly made no declaration of their intensions to leave us yet we could readily perceive a refusal would be productive of that effect.' They were given an increase of 8d per fodder.[49] In 1779 the price was raised to 7s a fodder, but in the next year it went down to 6s 8d again.[50] In 1802 the price was less than the 7s 6d paid at Nenthead, and the Langley rate was increased accordingly.[51] In 1830 the price was still 7s 6d a fodder for ore hearth smelting, and every other job in the mill had a rate laid down for it.[52]

For maximum efficiency the furnaces had to be kept going for long periods at a time, with periodic stops to allow them to cool down. Smelters were therefore expected to work long and erratic hours. At the ore hearth the normal duration of a shift was from twelve to fifteen hours.

At mills where the smelting shift is 12 hours, the hearths usually go on 12 hours, and are suspended 5. . . . The two men, who manage the hearth, each work four shifts per week; terminating their week's work at three o'clock on Wednesday afternoon. They are succeeded by two other men, who also work four 12 hours shifts; the last of which they finish at four o'clock on Saturday. In these eight shifts from 36 to 40 bings of ore are smelted, which, when of good quality, produce from 9 to 10 fodders of lead. At other mills where the shift is from fourteen to fifteen hours, the furnace is kindled at four o'clock in the morning, and worked until six or seven in the evening each day, six days in the week. . . . Two men at one hearth, in the early part of each week, work three such shifts, producing about 4 fodders of lead—two other men work each three shifts in the latter part of the week. . . . Almost at every smelting mill a different mode of working, in point of time and quantity, is pursued.[53]

This was in 1831, but references in the Greenwich Hospital papers show that mid-eighteenth-century conditions were much the same. Each roasting furnace in 1831 was generally worked by four men, in partnerships of two, 'each pair of men work 8 eight-hours-shifts per week, and are relieved at the end of each shift by another couple, who work with them alternately eight hours on and eight hours off'.[54] The slag hearths and refining furnaces were similarly worked in long shifts. The 1842 report gives other permutations of hours[55] but they are similar in that each man worked long shifts, followed by a long break before the next set of shifts.

The eighteenth-century smelt mills were probably most unhealthy. Lead poisoning from fumes was a common and fatal illness in the Scottish mills of the time.[56] No evidence on this point survives for the northern Pennines but it was in this region that 'horizontal chimneys' were first introduced at the beginning of the nineteenth century. Sometimes more than half a mile long, these chimneys ran up the fell sides, taking the fume away from the mills and their surrounding settlements. They were primarily intended to recover the vaporised lead which was deposited on the walls as the fume cooled, but they greatly improved ventilation. In 1842 the atmosphere of the smelt mills, although unpleasant, was less lethal than that of the mines.[57]

In the eighteenth and early nineteenth centuries smelting, like washing, was not a full-time occupation. The cessation of washing during the winter months and the impassability of the carrier ways meant that the supply of ore normally dried up before spring brought new supplies. Fuel also was often scarce in the winter, and severe frosts froze up the water wheels that drove the bellows. At Langley mill finding the men occupation during these periods of idleness was one of the mill agent's chief worries. In 1769 a local farmer was asked to employ the mill's refiners while the smelters were 'busy with the Unrefinable Ore'.[58] In 1797 the men were set to work to cut a drainage channel to the mill reservoir,[59] while in 1800 the agent was directed to 'take great pains to make all our Workmen good Hedgers and peaceable ones'.[60]

Transport

A vital part of the labour force was concerned with transport—of *ore* from the mines to the mills, and of *lead* from the mills to the market, which for most of the northern Pennines was Newcastle, although some Teesdale lead went to Stockton. The Mountainous terrain, the severe winters and the paucity of roads have already been dwelt upon. The business of transporting many tons of lead was a serious and difficult one. In the eighteenth century the Blacketts spent far more on transport than on smelting.[61] At the end of the century, and in the first half of the next, there was a revolution in the methods and organisation of transport.

In 1760 the roads were so abominable that packhorses or 'galloways' were the normal means of transport. Gradually, as roads were improved and new ones built, horse-drawn carts replaced most of the galloways. After some abortive canal schemes, the Newcastle & Carlisle Railway was built in the 1830's, followed, some years afterwards, by branch lines up most of the dales. Galloways remained in use as local transport until the end of lead mining but by 1870 the Beaumonts were using traction engines as well.

Only the pre-industrial forms of transport when great numbers of individual carriers were employed, are described here.[62] In the eighteenth-century Blackett/Beaumont and Greenwich Hospital letter books there is much detail about the carriers—far more than about any other branch of the lead mining labour force. The growth of efficiency in the transport of ore and lead is illustrated by the fact that in the eighteenth century there is letter after letter in both series of records concerning problems relating to the carriage; by the 1830's, when both series end, there is very little indeed.

Until the last years of the eighteenth century the ore and lead were carried out of the mining area purely by horses. In 1768 'The Ore is carried from the Mines to the Mills entirely on Horseback; Galloways being employed carrying two Pokes of Ore, each weighing 1 cwt., that is $\frac{1}{8}$ of a Bing, consequently a Bing is carried by 4 Galloways.'[46] By 1805 the ore was carried to Langley 'on galloways . . . or in small one-horse carts, containing a bing or 8 cwt. each.'[63] Bailey and Culley, writing on the agriculture of Northumberland and Cumberland about 1800, noted the use of these single-horse carts for carrying lead, and recommended their adoption for general agricultural use throughout the area.[64] The use of carts increased as new roads were built, but some galloways were still employed right up to 1880.

The administration of the ore carriage caused many problems. The weather prevented much being done during the winter—the packhorse tracks were impassable in snow or heavy frosts, and even heavy rain, channelled down the tracks, changed them into quagmires. In the summer, when the roads were at their best, the carriers subordinated the transport of ore and lead to the needs of the hay harvest.[65] They took their own time in carrying the lead. A Blackett letter of 1770 demanded that 'This practice of the carriers of getting Lead into their hands and then taking their own time of bringing it must be broke through if possible'.[66] In the same year, Mulcaster, the Langley mill agent, made a special trip to Hexham to 'make a particular enquiry whether there is any Lead left between Hexham and Newcastle'.[67] As a result of this inspection the Hospital introduced a ticket system; the carriers who picked up the lead from the mill to carry it to Hexham had to get it receipted by the carriers who took over from Hexham to Newcastle, or by the Newcastle agent if the lead were carried all the way.[68] By enforcing this system the Hospital knew which carrier was responsible for a given quantity of lead, even if it changed hands *en route* from Langley to Newcastle. A similar system was still in use both by the Hospital and the Beaumonts in the 1830's.[69]

The lead pigs frequently had to be stored or left for a time by the roadside. Cobbett noted between Newcastle and Hexham in 1832 'loads of these *pigs* lying by the road-side, as you see parcels of timber lying in Kent and Sussex, and other timber countries. No fear of their being stolen: their weight is their security'.[70] Singularly few, in fact, were stolen. Thefts are rarely mentioned in the records (a great fuss was made when they did occur). When it did happen, the theft was normally by one set of carriers stealing from another 'for the purpose of making up some deficiencies of the Lead they took from the Mill'.[71]

Most of the carriers lived outside the mining region proper. Although most mining families had smallholdings, poor soil and bad weather conditions would not allow many horses or ponies to be wintered in the area. The bulk of the carriers transporting lead from Alston Moor, Allendale and Weardale came from the farming lands near to, but off, the Pennines. Most of the Greenwich Hospital's carriers came from around Hexham and Corbridge; Hexham was the meeting place when the Hospital called a meeting of all its carriers in 1769, and it was there that pays took place. In 1822 a Greenwich Hospital report noted that farms in the Hospital's Alston manor (off Alston Moor itself) were held at above their value 'but being generally let to Persons who obtain part of their Livelihood by the carriage of Ore from the Mines to the Smelting Mills, they are thus enabled to maintain themselves'.[72] The Hospital provided lodgings for carriers in need of them at Langley mill.[73]

The local inhabitants did not always co-operate with the carriers. In 1791 the Blackett land agent in Hexham found it necessary to warn the inhabitants of Allendale that a new pound had been built near the town 'for the Convenience of the Inhabitants of Allendale but not for Oppressing . . . ore and Lead Carriers'. Ponies belonging to the carriers had been impounded, and released only on payment of a fine. If the practice did not cease 'the said Common Pound shall be pulled down.'[74]

The carriers were mainly small farmers, frequently tenants of the landowning lead mining concerns, carrying lead as a supplementary source of income, although by the end of the eighteenth century an increasing number were virtually full-time carriers, using their land only to graze their horses. There was a definite change of policy here on the part of the mining concerns. In 1771 the Blackett chief agent wrote that 'care must be taken that by giving any Sett of Men an Exclusive Right to carry the Lead we do not put it in their power to carry only when it suits their own convenience without regarding the want there may be of the Lead at Markett'.[75] Part-time labour, with frequent changes of route, was preferred, under the (mistaken) belief that this would avoid combinations among the carriers to force up the price of carriage. By 1795 on several of the routes from the different mills to Newcastle the carriage was undertaken by one sub-contractor, responsible for the organisation of all the carriers on his route. At that time the Rookhope contractor was proving unsatisfactorily slow, and he was threatened that if he did not speed up transport 'the Carriage from Rookhope must be made an open Carriage and advertised as such'.[76] The two systems went on side by side until most of the work was taken over by the railways.

The theoretical basis of the prices offered to the carriers for carrying lead a given distance was what the employer considered to be a fair sum for the distance involved, allowing for the price of corn for the animals. In 1808, for example, the Greenwich Hospital gave 5d per pig for lead carried from Langley to Hexham; 10d from Hexham to Throckley; 5d from Throckley to Newcastle—1s 8d for the journey from Langley to Newcastle. These prices were an advance on those previously offered, 'in consequence of the high price of oats'.[77] Earlier, in 1760, a Blackett agent wrote of a carrier, 'It is unreasonable in him to insist on the same price now, that was paid when Corn was double the price'.[78] Another consideration was the existence of 'back carriage'; the Greenwich Hospital paid its carriers more for carrying ore from Alston Moor to Langley mill in 1805 than the London Lead Company paid its carriers over an equivalent distance, as 'their [the company's] carriers derive considerable advantage in back carriage'.[63]

These, then, were the factors the companies considered legitimate bases for fixing the price of carriage. The ones that really counted— the number of carriers available, the demand by different mining companies for their labour at any one time, and the urgency or otherwise of getting lead to market—were admitted to exist, but it was considered very unfair practice if the carriers used them as arguments for higher rates. As a Blackett agent wrote in 1764, at a time of great competition for their services, it was to be hoped that 'they will not show themselves of a restless and imposing disposition'.[79] He was frequently disappointed!

Competition between the different concerns frequently became open. A price rise by one forced others to raise their prices. In 1764 'the Quakers Co. have this week agreed to give 3/9 for their ten ps. so we cannot expect to get Sir Wr's 12 ps. carried for less than 4/–'.[79] In 1771, however, the Blackett agent doubted whether prices should be raised, as 'the other proprietors will do the same and if they find we one Year get their Men from them they will try another Year to get the Advantage of us and so we shall have a fighting trade.'[80] By the end of the century, unofficial agreements had been reached between the companies not to fight one another, although they were broken from time to time when there was a desperate need to get lead to market.

In this situation—the different concerns in rivalry with one another, and the carriers working for the highest bidder—the unfortunate farmers whose landlords had mining interests were used as captive labour to keep transport prices down. In 1762 Blackett tenant farmers near Dukesfield were urged 'to pay their rents by the Carriage'.[81] Two years later the Blackett chief agent wrote to Sir Walter that he should not sell lands he owned around Winlaton, as 'Combinations among the Carriagemen for price' were often put an end to by 'your Tenants . . . moving first with the Lead'.[82] In 1796 there was a revealing alteration in the terms of letting two Beaumont farms at Parkhead. The farms had been let on a twenty-one year lease to a Mr Simpson, who had sublet them, 'obliging the Tenants to lead Coals from his Collieries. . . . The farmers, . . . obliged to lead Coals the whole year, commonly neglect their farms . . . and they are in general bad Husbandmen.' The lease having expired, 'I would recommend that the two Farms be advertized . . . and let to good tenants subject to . . . conditions . . . *particularly* that of not leading coals for *any* person and being obliged to lead Lead.'[83]

But there were not enough tenants for the big mining concerns to rely solely on their labour. Additional labour had to be obtained by bargaining in a free market. In the mid-eighteenth century— particularly in the decade following the establishment of Langley mill in 1759—the Blackett/Beaumont and Greenwich Hospital records are full of references to strife with the carriers. The carriers would combine, and refuse to take the lead unless prices were raised; the employers changed routes, forced their tenants to carry lead, and attempted to exclude troublesome groups and individuals from the carriage to keep costs down. Only when they were desperate would they agree to raise prices. The arts of industrial bargaining were employed on both sides. 'Do not be too Urgent with them,' wrote the Blacketts' chief agent in 1762, when there was a seller's market for lead, 'lest you occasion a demand for more price'.[84] But market conditions normally determined the victor of each contest: in times of high demand the mining companies were prepared to pay almost anything to get their product to Newcastle; at other times the obdurate carriers could be reduced to beggary by withholding the traffic. A few examples will suffice.

In 1757 the Blacketts' chief agent wrote, 'I am told the Fairlams insist on an advance price, and they only, if others will lead it at the old price, imploy them, and let the Fairlams mind their Farm. But if nobody will lead it without an advance we must be forced to comply, for the lead must be got to Markett with all expedition.'[85] In 1762 the route from Rookhope was changed to 'teach Mr. Parker and his Confederates that they have not such a property in Rookhope Lead as they imagine and that the higher they raise the price of the Carriage the less they will have of it to carry'.[81] In 1764, however, the Blacketts was forced to yield to the demands of the Rookhope carriers, 'For the great quantity that is there makes it necessary for Sir Walter to submit at this time.'[86] In 1770 demand was light, and the Greenwich Hospital receivers could afford to write to their agent at Langley, 'We are not now in any hurry at all and if another Pig does not come from the Mill this Season we are resolved not to alter the Terms.'[87]

In the late 1780's the comment on carriage problems begins to fall off. When they were mentioned, the problems were similar to those in the earlier period, but the owners were gradually gaining the upper hand. Organisation had improved, and a 'fighting trade' between the different companies was avoided. In the Greenwich Hospital correspondence there are frequent instructions that such and such a person should not be employed in the carriage again. For the smallholders who depended on carriage rather than farming for a livelihood such a prohibition was ruinous. In the correspondence of the clerk of the Hexham manorial court there are two letters, written by an unemployed carrier and his wife who had been dismissed for signing a petition in September and August of 1796, appealing to Mrs Beaumont for reinstatement. Otherwise 'it does entirely reduce us to great hardship and will be a means of bringing us into oppressing Want for the situation of our Countrey Depends upon Employment of some kind or other that arising from your lead mines to earn Wages by to procure for ourselves and our family's amentainance'.

Crises occurred on relatively few occasions, compared with the mid-eighteenth century. In 1809 after a period of depressed prices when it was scarcely worth transporting lead from stores at the smelting mills to similar stores in Newcastle (the smaller companies simply could not afford to do so), prices suddenly went up. The Beaumont agent wrote that, in spite of his increasing the carriage rates, insufficient carriers could be obtained, 'as the Alston Moor Mine proprietors have more extensively advanced theirs and in all probability would continue to do so if Col. Beaumont's were again to be advanced, as many of these Miners are under the necessity of having their Lead Ore brought to market to procure money to carry on their Mines and therefore will give any price to secure the Carriers'.[88]

But the carriers' power to oppose the big mining companies was disappearing fast by the early nineteenth century, and had disappeared altogether by 1830. The new roads allowed the carriage to continue for most of the year; the carts now employed were quicker and required fewer men and horses then the strings of galloways. The coming of the railways greatly reduced the distances that had to be covered by non-mechanised transport.

The agents

In the larger mining concerns the complex operations and the numbers of men employed required organisation and supervision. Those responsible were commonly known as 'agents'; very junior grades were simply 'clerks,' 'inspectors of miners' or, in the nineteenth century, 'overmen'; also, in the nineteenth century, the chief agents of the London Lead Company and the Beaumont concern were known as 'superintendents'.

The surviving evidence about the duties and remuneration of agents is limited, as usual, to the largest concerns, the Blackett/Beaumonts, the London Lead Company and the Greenwich Hospital. Enough survives to show a clear difference between eighteenth- and nineteenth-century conditions. In the earlier period the number of agents was relatively low, their duties were vague, their wages poor and subsidised by forms of truck; in the latter their numbers relative to the number of men increased with the appointment of more low-grade overmen, their duties were more clearly defined, their wages improved, and truck vanished. In the London Lead Company the change came after Robert Stagg became superintendent in 1816; in the Beaumont mines, reform was slower. Although starting at the beginning of the century, it was not complete until after Thomas Sopwith's appointment in 1845.

In the Blackett/Beaumont mines the eighteenth-century structure of administration was as follows. In Newcastle the chief agent, assisted by clerks, was responsible for the general control of all mining, smelting and transport operations, and for the sale of lead. Each of the three major districts—Weardale, East and West Allendale—was under the control of an agent, as was each of the smelting and refining mills. Each mine agent had a number of assistant agents, responsible for book-keeping, general inspection of the mines, and washing operations.

In 1805 a report on 'the future Agency and Management of their Lead Mines' was presented to Colonel and Mrs Beaumont, detailing the numbers, duties, wages and privileges of the agents. The principal agent of each district took 'upon himself the *whole responsibility*, consequently consider himself the inspector of *every department* within his district.' He was assisted by a clerk and a washing agent, and in Weardale also by an assistant agent. The central recommendation of the report was the creation of a new post to be known as 'Inspector of Miners'—one general inspector, together with one local inspector in each district, two in Weardale. These inspectors were to be responsible for the day-to-day management of the mines. 'It will be required that such Inspector go into all the Mines within his district every day, and take particular care that all the working Miners do their duty.' He had to be 'capable of taking levels, hatching and laying down correct plans of the different workings, not merely describing the progress of such workings, but also the dead work done at each place'. He was to be responsible for keeping levels clear, for supplying materials needed and for informing the principal agent 'of all complaints, neglects or any other matter of which he does not approve'. This principle of tighter local control was almost revolutionary, but it was effectually negated by the small number of inspectors appointed. Daily inspection of every working was quite impracticable.

The appointment of Thomas Sopwith as chief agent in 1845 brought about a more substantial revolution. He was resident in the area, with his office not in Newcastle but at Allenheads. He commented on the situation, as he found it, in his diary (unfortunately very discreetly). The structure of the administration led to 'ruin and decay as regards material objects and to disunion and ill-feeling in the minds of all parties concerned. . . . Three distinct and rival dynasties had grown up in the valley of the Wear and the two dales . . . of Allendale'.[89] Sopwith ended the supremacy of the district agents. His diaries show that his rule was absolute, almost every decision being referred to him. The vast improvement in communications that had taken place in the early nineteenth century made this possible. He also increased the number of inspectors, now called overmen, to supervise every aspect of underground work. In 1864 every working was visited at least twice a week.

The London Lead Company had never been so loosely organised as the Blackett/Beaumont concern in the eighteenth century. Executive control over the whole of its operations in the northern Pennines was in the hands of one agent, whose headquarters were first at Nenthead and then at Middleton. The district agents were responsible to him. Robert Stagg's most important administrative reform was the introduction of overmen to provide local control over all operations. In a report of about 1820 it was stated that 'the most important improvement made in the management of the concern was the introduction of *Mine Overmen*, taken from the ranks. Prior thereto, the mine workings were visited by an Agent very rarely, leaving the miners to play tricks, and take advantage of all kinds; but the daily and hourly inspection of overmen puts a final close to all such, with an improved economy which scarcely admits of being fully estimated'.[90]

Promotion to agent was a progress by a blend of corruption and merit. A worthy man was far more likely to achieve promotion in the nineteenth century (if only because of the greater number of openings) than in the eighteenth, although even in the earlier period he would not have found it impossible. In the two largest companies the agencies were all limited to local men with relevant experience. On several occasions in the Blackett/Beaumont correspondence there is discussion about the appointment of a new agent or sub-agent. Throughout the eighteenth and nineteenth centuries it was felt that 'It is not at all desirable to introduce Strangers into this Establishment'.[91] When Colonel Beaumont suggested in 1820 that a Cornishman might be employed, the chief agent ponderously replied that he was convinced that 'your Mine agents are in ability well qualified for their situation, and highly respectable for integrity of Character; should however you and Mrs. Beaumont conclude that a person from the Cornish mines strongly recommended, altho' the produce of lead there is comparatively small, is more competent to conduct with greater advantage Mines in Northumberland and Durham in all respects different and of which he can have no practical (the most essential) knowledge, it will then be a subject for your determination, and how far the change, from a system hitherto successful, may in its application affect important interests at stake, is a consideration which deserves and undoubtedly will engage your most serious attention.'[92] The Cornishman did not get the job!

All the mining companies normally preferred to promote men to the lower agencies from within their own ranks. Each Beaumont inspector of Miners of 1805 was 'to be selected from the working Miners at one of those Mines over which he is not to be employed as Inspector.' More senior posts were often filled by sub-agents from other companies. William Westgarth, the inventor of the water pressure engine at Coalcleugh, had been a Lead Company sub-agent before he was appointed to Coalcleugh. The Blackett/Beaumont chief agent was sometimes promoted from within the concern—as were William Crawhall in the early nineteenth century and J. C. Cain later—or brought in from outside—as was Thomas Sopwith.

Meritorious conduct and suitable experience were the reasons urged when an individual was promoted, but the numbers of members of the same family working as agents for one company show that family connections played an important part. The Crawhall family, for example, dominated the Beaumont mines from about 1810 to 1845. The brothers William and George Crawhall were agents of East Allendale and Weardale respectively and William subsequently became chief agent. Their father had been Allenheads agent, too. The

Stagg family enjoyed a similar status in the London Lead Company from the end of the eighteenth century to 1843.[93] And members of the Mulcaster family appear to have held nearly every smelting agency in the northern Pennines. When Peter Mulcaster was appointed as the first agent of Langley mill in 1768, two of his brothers held similar positions under Sir Walter Blackett and the London Lead Company.[94] The Langley agency was held in succession by different members of the family until after 1830.

By 1864 promotion by merit rather than by nepotism was established in both the largest mining concerns. Robert Bainbridge, the Lead Company's chief agent, told the 1864 commission how overmen and agents were appointed:

I apply to all the local agents to give me a list from time to time . . . of all the young men in the concern whom they conceive likely to make good inspectors. When I have obtained that list from them I keep my eye upon each of these parties, individually, and form my own opinion of their character; and our test, of course, is intelligence, integrity, and independence of character.[95]

William Curry, the Beaumont agent at Allenheads, was asked by the commission for a sketch of his career. He had begun work about 1820, aged 11 or 12, blowing air machines. Later he became 'a hireling' of the bargain workers before he was admitted into bargains as an equal partner. He was ten years a miner, and had been twenty-seven years an agent of one rank or another.[96] In 1860 Sopwith jotted down in his diary a history of the career of the oldest Beaumont agent then in service—Mr Steel, aged 77. He had spent sixty-one years of his life (i.e. beginning in 1799) in smelt mills. He had worked as an ordinary smelter at Dukesfield mill for nineteen years: in Rookhope mill for seventeen, first as a workman, then as a clerk, and finally seven years as sub-agent. He had then gone to Allen mill as agent (succeeding H. L. Pattinson) and had been there twenty-five years.[97]

The agents' salaries, particularly in the eighteenth century, were not so very much above what a reasonably successful miner would expect to make in the course of a year, but they were reliable and constant (unlike the miner's earnings), and there were extraneous forms of remuneration.

In 1760 a Blackett mill clerk, i.e. deputy to the mill agent, was paid £20 a year, together with a 'little Farm and House which belongs to the Office'.[98] In 1768 the Greenwich Hospital appointed Peter Mulcaster as agent at the new smelting mill at Langley; he was to receive the same salary and conditions as the London Lead Company gave its Nenthead mill agent—'£30 per year, an House to live in, and Coals for Fire; they generally have two or three acres of land at moderate price to keep a cow'.[99] The new second agent at Allenheads mine in 1777 was paid £30; his predecessor had received £40 and both of them were charged £11 15s for the small farm they rented.[100] At the beginning of the nineteenth century the mine agents of Weardale and Allenheads received £100 a year. The agent in charge of the smaller Coalcleugh area was paid only £60. The assistant agents were paid £50 or £60 a year (see Appendix 2).

In the nineteenth century the London Lead Company's agents were considerably better paid than those working for the Beaumonts. From 1817 onwards the company's chief agent was paid £1,000 a year;[101] in 1865 Thomas Sopwith's salary for his equivalent post in the Beaumont concern was only £300, which had become £600 by 1870. In 1857 the company's overmen were paid on a progressive salary scale, with a maximum of £80 a year; in 1865 this maximum had been increased to £100.[102] Overmen in the Beaumonts' employ were paid only about £50 in 1857. An additional advantage for the Lead Company's employees was that the company made a habit of paying them gratuities when lead was selling well.[103]

The remuneration of agents cannot be compared without taking into account the 'fringe benefits' they received. The most important has already been mentioned—a house and farm rent-free, or at a reduced rent. Sopwith had a house built for him at Allenheads at the owners' expense as a condition of his accepting the post of chief agent. Forms of truck (see chapter 11) died out during the nineteenth century but more quickly in the London Lead Company than in the Beaumont mines.

The nineteenth-century increase in the number of agents was a vital part of the modernisation of the bargain system. Sub-contract continued, but it was closely supervised. Detailed planning and control of all mining operations at last became possible.

NOTES

[1] Jars, vol. 2, p. 545: ' . . . but with still less care; so that they lose a quantity of the ore, apart from that which is deliberately thrown away.' Mulcaster MS.

[2] Blackett of Wylam MSS.

[3] Sopwith, 1833, p. 120.

[4] Forster, 1821, p. 351.

[5] P. A. Dufrenoy and J. B. A. L. L. Elié de Beaumont, *Voyage métallurgique*, second edition, 1837–39.

[6] G.H. (Wigan) MS.

[7] L.L.C. 16a, 13 July 1811.

[8] B/B 54, Lady Day 1829.

[9] B/B 50, 18 March 1797.

[10] B/B 53, 20 October 1809.

[11] Jars, vol 2, p. 543.

[12] B/B 136, 29 March 1779.

[13] L.L.C. 16a, 18 January 1812.

[14] L.L.C. 16a, 1 November 1811.

[15] Forster, 1821, p. 363.

[16] B/B 167, 14 October 1766.

[17] 1842 report (Mitchell), pp. 763–4.

[18] Ibid, p. 746.

[19] T. Sopwith, jr, 'The dressing of lead ores'. *Minutes of the proceedings of the Institute of Civil Engineers*, vol. 30, 1869–70, part 2, p. 108.

[20] 1842 report (Mitchell), p. 760.

[21] Ibid, p. 768.

[22] Ibid, p. 766.

[23] Ibid, p. 732.

[24] 1864 report (vol. 2), p. 331.

[25] Raistrick, 1938, p. 38.

[26] These included the parishes of Alston, Middleton in Teesdale and Stanhope.

[27] 1842 report (Mitchell), p. 747.

[28] 1864 report (vol. 2), p. 376.

[29] 1842 report (Mitchell), p. 743.

[30] Ibid (Leifchild), p. 682.

[31] Ibid (Mitchell), p. 745.

[32] For a list of smelt mills in the northern Pennines, see A. Raistrick, 'Lead smelting', *Transactions of the University of Durham Philosophical Society*, IX, 1931–37, pp. 164–79.) Dr Raistrick's list is far from complete, particularly for eighteenth-century activity.

[33] This account is based on: Mulcaster MS; H. L. Pattinson 'Account of the method of smelting lead ore in . . . Northumberland, Cumberland and Durham,' *Transactions of the Natural History Society of Northumberland, Durham and Newcastle upon Tyne*, 2, 1831, pp. 152–77; J. Percy, *The metallurgy of lead*, 1870.

[34] Greenwich Hospital report, 1832, pp. 15–16.

[35] Pattinson, pp. 152–77.

I

[36] B/B B.M. letter book, 10 June 1730.

[37] Adm. 66–96, p. 65.

[38] Mulcaster, MS.

[39] Raistrick, 1938, p. 69.

[40] Adm. 66–96, p. 117.

[41] Adm. 66–96, p. 179.

[42] Adm. 66–96, p. 312.

[43] Adm. 66–97.

[44] Adm. 66–96, p. 164.

[45] Adm. 66–100, p. 188.

[46] Adm. 66–96, 20 April 1768.

[47] Adm. 66–96, p. 208.

[48] Adm. 66–96, p. 255.

[49] Adm. 66–96, p. 305.

[50] Adm. 66–97, 22 June 1779; 9 June 1780.

[51] Adm. 66–100, 21 August 1802.

[52] Adm. 66–103, 23 February 1830.

[53] Pattinson, pp. 161–2.

[54] Ibid, pp. 154–5.

[55] 1842 report (Mitchell), p. 735.

[56] T. C. Smout, 'Lead mining in Scotland' in P. L. Payne, *Studies in Scottish business history*, 1967, p. 127.

[57] 1842 report (Mitchell), p. 736.

[58] Adm. 66–96, p. 55.

[59] Adm. 66–99, 28 August 1797.

[60] Adm. 66–100, p. 89.

[61] Hughes, p. 72.

[62] For an account of overall transport development, see A. Raistrick and B. Jennings, *History of lead mining*, 1965, pp. 267–76.

[63] Greenwich Hospital report, 1805, pp. 49–50.

[64] J. Bailey and G. Culley, *Agriculture of Northumberland*, 1813, p. 273.

[65] B/B B.M. letter book, 28 May 1731, a letter saying that as much carriage as possible should be completed before the hay harvest.

[66] B/B 48, 8 July 1770.

[67] Adm. 66–96, p. 72.

[68] Adm. 66–96, p. 75.

[69] B/B 52, 5 September 1833.

[70] W. Cobbett, *Tour in Scotland*, 1833, p. 62.

[71] B/B 50, March 25, 1797.

[72] Greenwich Hospital report, 1822, p. 2.

[73] Adm. 66–103, 3 January 1826.

[74] Hexham manorial papers.

[75] B/B 48, 22 January 1771.

[76] B/B 50, 9 August 1795.
[77] Adm. 66–101, 29 October 1808.
[78] B/B 47, 11 April 1760.
[79] B/B 48, 11 May 1764.
[80] B/B 48, 5 March 1771.
[81] B/B 47, 26 July 1762.
[82] B/B 48, 15 February 1764.
[83] B/B 50, 22 March 1796.
[84] B/B 47, 10 September 1762.
[85] B/B 46, 13 April 1757.
[86] B/B 48, 29 June 1764.
[87] Adm. 66–96, p. 78.
[88] B/B 53, 30 June 1809.
[89] Sopwith, *Diary*, 15 February 1861.
[90] Quoted by Raistrick, 1938, pp. 72–3.
[91] B/B 51, 7 April 1809.
[92] B/B 51, 21 November 1820.
[93] Raistrick, 1938, pp. 141–7.
[94] Adm. 66–96, 20 April 1768, p. 4.
[95] 1864 report (vol. 2), p. 380.
[96] Ibid, p. 367.
[97] Sopwith, *Diary*, 30 April 1860.
[98] B/B 47, 9 January 1760.
[99] Adm. 66–96.
[100] B/B 49, 9 February 1771.
[101] Raistrick, 1938, p. 144.
[102] L.L.C. 31, 5 September 1865.
[103] Raistrick, 1938, p. 54.

VI

Strikes and industrial disturbances

The incomplete nature of the evidence makes it particularly difficult
to set down the history of labour relations in the lead industry. About
many disputes no details survive; when evidence does exist it is nearly
always from the side of the masters.

The overall strike record

There is no record of any strike or combination until the last decade of
the eighteenth century, the one exception being the carriers, whose
joint efforts to secure higher rates have just been described. In other
branches of the industry workmen apparently did not normally strike,
though they sometimes rioted or sabotaged their employers' property.
Surviving letter books of the Blackett concern cover various periods of
crisis in the eighteenth century when bargain prices, or the size of the
labour force, were cut back. In 1731 the chief agent was 'almost
pulled in pieces in Weardale' by men whose pay was some two years in
arrear. In 1760 after a 'tumult' at Allenheads, the chief agent ordered
the ringleader to be discharged and others to be cautioned, 'for Sir
Walter is determined to have good order and discipline kept up'.[1] In
1775 the bellows at Langley mill were 'every one of them, cut . . . It
is not unlikely but that the Mischief has been done by some of the
Workmen who have been discharged from the Hospital's Service on
account of their ill-behaviour or want of abilities.'[2] There is no record
of any strike, as such, before 1796.

The apparent absence of strikes *may* be due to lack of evidence, but
mining in the eighteenth century was organised very much on a basis
of contract between free parties, rather than between employer and
employed. Strikes were unlikely in such circumstances. In the 1790's,
a time of war, rising prices, shortages and much unrest throughout
England, a strike did occur in Weardale, caused mainly by discontent
over the failure of bargain prices and subsistence payments to keep pace

with the rapidly rising cost of living. This strike, together with the events preceding and succeeding it, will be described later. From references in the Beaumont records it seems there were disturbances among the London Lead Company's miners about the same time. In 1795 a number of Lead Company men were discharged for taking part in a strike.[3] In 1797 there was a riot on Alston Moor, when 'about 200' miners met and 'proposed to lay that part of the country under Contribution and had named the persons and the sums they insisted on having'. They dispersed after being addressed by the Lead Company's agents.[4]

In the nineteenth century the evidence is more complete, and there is reasonable assurance that all strikes in the major concerns are recorded in some way, even if no details are known. It will be convenient to take the two large companies separately.

The London Lead Company suffered only two actual strikes during the whole of its mining history in the nineteenth century. There was a minor incident in 1811, when the agent at Boltshaw mine was 'obliged to turn away all the workmen this quarter end and send fresh hands from Nenthead. Such a parcel of designing Men I have seldom met with'.[5] In 1816–17, when Robert Stagg first took over the chief agency of the mines, there was apparently a strike, about which no details are known, against his methods of retrenchment to deal with the low prices of lead that year.[6] From then on, including the difficult years of the early 1830's, there was none until 1872, when a 'prolonged strike' occurred in Teesdale. It is not mentioned by Dr Raistrick, but the *Mining Journal* for 1876 records that subsistence money was much increased as a result of it.

Unfortunately these few facts are all that is known about industrial disputes in the Lead Company; there is also a certain amount of evidence, discussed more fully in chapter XI, illustrating the company's paternalism. It was to the strictness of the moral rules enforced on the miners that Stagg in 1847 'attributed the total absence of rebellions and insubordinations of every kind; and that chartism—radicalism—and every other abomination have for so many years been strangers to the concern'.[7] Trade unions would not have been welcome in such an environment.

The Beaumont concern was nearly as free from strikes as the London company. Only three major ones occurred in the century, each restricted to a particular district. In 1819 there was a strike in Weardale, the climax of some years of discontent. It will be described in detail shortly. In 1849 there was a strike lasting four months at Allenheads. Fortunately a great deal of evidence has survived concerning this interesting dispute, the cause of which was Sopwith's efforts to transform the miners from independent sub-contractors into employees bound to work regular hours. This too is described later. Finally, the last years were bedevilled by a series of strikes in Weardale over the reduction of the labour force and the new bargain system introduced by the unpopular chief agent, J. C. Cain. In this case, practically nothing survives to record the course of events. The strikes failed, for the owners could only save money by the suspension of work, so disastrously had the market collapsed.[8] The Beaumonts abandoned their Weardale leases in 1883.

Three strikes are reasonably well documented: Weardale, in 1795–96; Weardale, 1818; and Allenheads, 1849.

Weardale, 1795–96

It is obvious from the tone of the letters in the Beaumont letter book from 1794 to 1800 that the agents were very much aware of discontent among the workmen. The period was one of country-wide depression and rising food prices. The price of lead fell at the beginning of the decade, rose in 1795–97 and then dropped again. In December 1794 'some papers . . . of a riotous and seditious tendency' were in circulation at Allenheads, and numbers of miners petitioned for subsidised supplies of rye.[9] They were supplied, but discontent continued.

In November 1795 some lead miners from Rookhope seized flour from unpopular millers, and the chief agent received a deputation from Weardale with a petition signed by about 500 workmen who had stopped work at their bargains.

They request that their Wages may be raised, and their Subsistence Money every two months doubled. I had a good deal of discourse with them, and they were very sensible that you had done more for the relief of your Workmen than any other Proprietor etc. . . . We thought it advisable *at this time* to increase their subsistence Money one half, to continue the supply of Corn at a reduced price . . . and to increase bargain prices. The Miners left me *well satisfied,* and . . . they promised me that the whole of them would be at work tomorrow, or next day.[10]

The men returned to work, but the next month, December, the chief agent noted the presence of 'a few Refractory Men in Weardale who

I have directed the Agent to discharge'.[11] These 'dissatisfied turbulent men' were urging the other miners to stand out for double subsistence money.

In 1796 corn prices continued very high, and the discontent in Weardale did not die down. In April the chief agent again threatened to discharge 'mischevious people' who were causing 'disturbance'.[12] Despite this threat there was another stoppage in August. Unfortunately the details have not survived—there is a gap in the chief agent's letter book at this time. What has survived, however, is the petition signed by some 180 Weardale miners (Appendix 3), which throws light on many aspects of their relationship with their employers. The main demands were for higher bargain prices and a further increase in subsistence money, but the petition is permeated by dislike of the agents—both the local Weardale agents and the chief agent, who, the miners believed, had 'intercepted' their November petition to Colonel Beaumont. The agents were accused of making excessive profits from horse hire, etc., and the final request in the petition was for 'such treatment as Rational beings are entitled to, which they have so long, so often but in vain expected.'

The results, if there were any, are not recorded. Subsistence was increased, but not till the following year. In January 1797 the chief agent penned a long letter to Colonel Beaumont defending the Weardale agents from charges in a 'Memorial of Ralph Coulthard and John Kidd, two of the Workmen who from my own knowledge have been the leaders in every disturbance amongst the Workmen of Weardale Mines'. The story of their getting vast sums from hiring horses to work in the mines was, he protested, 'a great falsity'. One of the younger agents had been 'a little sharp with some of the Men for which I found fault with him, and he promised me to be more circumspect and cautious for the future.' He urged Colonel Beaumont not to listen to 'these idle complaints which must give encouragement to these discontented fellows and weaken the authority of your Agents'.[13]

After this there was apparently no further trouble in Weardale for a while. The lead trade was more prosperous; bargain prices and subsistence payments both improved. The disputes of the mid-1790's were not forgotten, however, and the miners petitioned Colonel Beaumont whenever they had cause for complaint. The chief agent relieved his feelings in a letter to the Weardale agent in December 1799, after they had complained to Colonel Beaumont about corn prices. 'I have always found the Weardale miners ... the most dissatisfied, unreasonable and Turbulent Set of Workmen in the Concern.'[14]

These 'incidents' took place against a background of troubled times, but it is noticeable that discontent, or at least active discontent, was not spread throughout the whole Beaumont concern. There was no united front between the miners of Weardale and those of Allenheads and Coalcleugh. The agent's letters presuppose a militant group in Weardale inciting the others to take action. The petition is headed as from 'the Commetee of Weardale Miners'. The burden of the miners' complaints was that the price of food was rising faster than earnings. The specific complaints against the agents suggest that the area was tactlessly managed. Lastly, it should be observed that the miners were largely successful in their agitation—subsidised food was sent into the area, bargain prices were raised considerably, subsistence payments trebled in the decade, and the chief agent felt bound to tell his subordinates to treat the miners with greater tact and discretion.

Weardale, 1818

The second of the three strikes took place at a time when the fear of revolution was widespread throughout the country. As before, there had been an acute depression in trade. December 1816 saw the monthly subsistence paid to the Beaumont miners cut from 30s to 20s, and bargain prices reduced correspondingly. The chief agent and Colonel Beaumont were petitioned, and a delegation from Weardale visited the colonel at his Yorkshire home. The agent was against resuming subsidised supplies of rye to the area; on this score he was overruled by his master, but no advance was to be made in payments. The petitioning miners threatened not to take their bargains unless higher prices were offered.

I told them that under the present depressed state of the Lead Trade they could not be doing their employers a greater benefit than desisting to raise Ore for three Months to come and the only regret that would arise by their so doing would be the distress they themselves would feel, and the impoverishment which it would produce to the Country. This observation appeared to surprise and stagger them.[15]

In the same letter the chief agent complained that the London Lead Company 'by the introduction of new measures into their concerns have caused all this discontent among the Men'.

The miners, then, gained nothing from their threat to strike except resumed supplies of rye. There was much truth in the chief agent's observation that they were in an impossible bargaining position at the very nadir of the depression, since the loss of labour would be no worry to their employers.

There was no more trouble until the end of 1818, by which time the state of the lead trade had greatly improved. The 30s subsistence allowance had already been reintroduced when in September 1818 there was a renewed series of petitions and delegations from the Weardale miners demanding an increase to 40s a month, the rate the London Lead Company was paying. The petition of 24 September to the chief agent is reproduced in Appendix 4. Ten shillings per bing extra for raising ore was requested, and 40s per month subsistence, on the grounds that 'Lead has considerably advanced' but the rise in its value had not been passed on to the miners. The chief agent told the delegation which presented the petition that 'the present poverty of the Mines would not allow . . . the relief they sought for'. The miners 'expressed their determination not to Work unless their demand is complied with'.[16]

A second petition, enclosed with a letter from the chief agent to Mrs Beaumont at the beginning of October, reiterated the strikers' demands—40s a month and an increase in bargain prices. Unless they were granted, 'it will be impossible for the greatest number of us to get the necessaries of life, as our Credit is utterly gone.' In his accompanying letter the agent told Mrs Beaumont that he had just been to Weardale 'at the solicitation of Geo. Crawhall [the resident agent] who became alarmed by the menacing tumultous proceedings, of the largest assembly of Miners ever before seen there.' He met 'six of the Men' delegated by the others. They demanded '10/– per Bing advance and 40/– per month Subsistence Money'. The agent offered them 5s per bing advance. The concession was rejected and the chief agent left Weardale, pausing only to offer 'every protection' to men who chose to come to work, and to sign a letter informing local magistrates of the strike.[17]

The affair had now grown to such formidable proportions that news of it found its way into a confidential report, dated 6 October, from the mayor of Newcastle to the Home Secretary. The mayor recorded that 'the men have formed themselves into a large body controlled by what they call a committee, and after having published some inflamatory placards exciting to tumult, have threatened not only to

desert their engagements and their work, but to do damage to the
mines and to the works if their demands be not complied with'.[18]

Writing to the Weardale agent on 9 October, the chief agent
opined that although they were still holding out for their full demands,
'the Men will return to their Work soon and that an *apparent* in-
difference on your part to their proceedings may have a good effect'.[19]
On 15 October it was confirmed that the miners had returned to
work, 'and the last quarter having expired for which the Ore Bargains
were taken, it becomes necessary to enter into a new agreement with
them'. Five shillings more per bing was paid at these new bargains.[20]
The mayor reported the end of the strike to the Home Secretary,
adding that the miners were given subsistence money for the period
they had been on strike.[21]

In a letter of 24 October the chief agent mentioned to Mrs Beaumont (unfortunately, full details appear to have been given in a letter
that has not survived) the 'proceedings of the Weardale Miners at
Allenh. and Coalcl.' These 'proceedings' apparently took place after
the settlement of the strike, so not all the strikers can have returned to
work. He suggested that information should be laid before a magistrate
against 'Geo. Robinson and Stephenson for obstructing those who are
inclined to work'. He doubted, however, whether he could procure 'a
sufficient number of special Constables to act with effect in Weardale'.[22] A letter to the Weardale agent said that Colonel Beaumont
was 'willing to pass over the offence [of the strike] in the hope that
similar disgraceful scenes will never occur again, yet it is with the
exception of those persons who have been the chief instigators in the
late transactions who are not to be employed again in his Works
whenever you can discover or identify those persons'.[23]

The strike was now presumably over, as there is no further mention
of trouble. Unlike the events of 1795–96, the 1818 strike ended in
almost complete failure for the miners. They had struck for similar
reasons—more subsistence and higher bargain prices—at a similar
time—when lead was rising in value after a depression. True they
obtained an extra 5s a bing, but this was a long way short of their
demands, and the outcome was capitulation on the owners' terms, with
the subsequent dismissal of 'the chief instigators in the late transactions', who presumably included most of the 'committee' mentioned
by the mayor of Newcastle. As in 1795–96, only the Weardale district was involved. The miners of Allenheads and Coalcleugh were
appealed to, but only after most of the strikers had surrendered.

Allenheads, 1849

The strike at Allenheads was rather different from the other two. It took place at a reasonably prosperous time when the miners were not greatly oppressed by generally adverse economic conditions. It was a reaction against modernisation, against changes introduced by a dynamic new chief agent that limited the traditional freedom of the miners, rather than a demand for higher wages. The new agent was Thomas Sopwith, an able engineer and geologist, and an enthusiast for social and educational reform. He mixed these virtues with a heavy Victorian pomposity, an overwhelming belief in the rightness of all his actions, and complete inability to see anyone else's point of view. He became chief agent in 1845, with specific instructions from the ageing owner, T. W. Beaumont, to carry out what reforms he considered necessary to increase the productivity of the concern.

There are three main sources of evidence regarding the 1849 strike. First, there is Sopwith's diary. Most unfortunately, the original one for 1849, written up daily, has not survived, and we are left with only a shortened version compiled in 1877, a few years before his death. Although the main facts are there, the drama has gone and Sopwith, who was inclined to sentimentality at the end of his life, may well have toned down some of the more distressing incidents. Second, there is an extremely interesting series of letters to the *Newcastle Guardian*. Some three months after the beginning of the strike a letter appeared in the issue of 24 March, signed 'The Allenheads Miners' and stating that as 'all available avenues of private communication have been closed against us, we are compelled to adopt this course as our last resort; and we trust that the decision of a discerning public, whose frown is ever the most potent weapon against tyranny and injustice, will so influence the minds of our oppressors as to lead them to restore us to the privileges of which we have been unjustly deprived'. The following week there was an answer to this letter signed 'T. Sopwith' and then followed four further communications from the striking miners which continued after most of the men had returned to work—the last two letters being signed 'The Ex-Miners of Allenheads'. Lastly, memories of the strike were still green in Allendale fifty years later when George Dickinson wrote his history *Allendale and Whitfield* (second edition, 1903) and he recounts a number of interesting anecdotes.

The history of the strike is as follows. Shortly before Sopwith took over the agency, a time clause had been introduced into most bargain contracts: 'to work five eight hour shifts per week per man'. Efforts to enforce it apparently caused much trouble, and Sopwith much unpopularity, during 1847. At the end of 1848 a 'spy system' (as the strikers called it) was introduced at Allenheads—sub-agents checked the time of entry and exit of all the miners to and from the mines. On 28 October 'a meeting of the miners was held, to consider and trace, if possible, the origin of this unprecedented treatment, and to adopt means for its removal'. A petition was sent to Sopwith complaining that the practice was unjustified and that the inspectors concerned were exercising 'powers which they had hitherto not possessed, to invade the acknowledged privileges of the men, and to seize every possibility of causing annoyance.'

Sopwith returned no answer to the petition and no enquiry was

made into the truth or falsity of the charges made against the inspectors. . . . He immediately ordered his name to be withdrawn from the Loyal Miners' Association [a local benefit society] declaring that he would not be associated with such a mobbish body; he denigrated the miners with all manner of opprobrious insults, as ignorant, indiscreet, Irish ruffians mobs &c, and even went so far in his rage as to obtain, through his land agent, the immediate discharge from their houses of four of the deputation. He also subsequently discharged from their employment fourteen who had the hardihood to prefer their charges against the inspectors.[24]

Sopwith, it may be added, in his single letter to the *Newcastle Guardian*, justified his not listening to the miners' petition by alleging that it demanded 'not an investigation into any alleged cause of complaint, but the dismissal of two of the under agents'. He went on to say, 'I consider the minds of the great bulk of the Allenheads miners to be inflamed by the ignorant and malicious conduct of a few demagogues, towards whom . . . I used the authority vested in me, to dispence with their future services.'

The miners appealed to a local magistrate, a friend of T. W. Beaumont, who was ill in Yorkshire at the time (he died in December). A compromise was reached whereby every miner signed a declaration that he would work an eight-hour day, and the watchers were withdrawn. An enquiry was promised into the causes of complaint against the sub-agents. However, the dismissed men were not reinstated, and so obviously had nothing to lose by upsetting the settlement. Sopwith soon gave them their opportunity. A week after the watchers were withdrawn, 'instead of the usual two, five time takers were present,

book in hand, and noting the hours of our arrival and departure; together with a constable duly armed with staff, pistols, and handcuffs.
And thus it continued till Christmas.' As for the enquiry, the miners
alleged that they were promised an enquiry under the chairmanship of
an independent outsider. Instead, when they went to the arranged
meeting, 'conceive our surprise at seeing our friend, Mr Sopwith, with
praiseworthy pertinacity, occupying the chair, and supported right and
left, by—nobody. After a restriction that the old charges should not be
entered upon (though that was the real object of the meeting) some
minor topics were introduced, but the chairman soon found a pretext
for dissolving the meeting, and thus nothing was accomplished.' The
new bargains, beginning in January 1849, were then due. 'None of the
400 men (except four, and these from peculiar circumstances) would
take bargains until the questions in dispute were settled.'

So battle was joined. As Sopwith remarked in his diary it was not
formally a strike. 'Strictly speaking it took the form of a refusal to
enter into a new contract for the first quarter of the year. This of
course was perfectly within the right of the Miners as free Agents
whether to work or not.' Sopwith had reintroduced the 'watching', he
said, because in spite of the agreement of 17 November, 'no improvement was attempted. The entrance to the Mines ... being in view
from my Office Window I had constant proof that instead of working
8 hours many miners only worked for 7 hours.' But Sopwith had
another reason for being as ready to accept a strike as the miners' dismissed leaders, a reason he did not express in his letter to the *Newcastle
Guardian* but which is mentioned in his diary: 'there were many more
miners at Allenheads than were required by the condition of the
Mines.' One would be hesitant to suggest that Sopwith precipitated
a strike as an excuse for trimming the labour force, but it was obviously
a factor he was conscious of.

There was no violence, nor any threat of it. In the early days there
was some negotiation between the contending parties, but both sides
were immovable in their attitudes. Sopwith noted his discussions with
a delegation of miners on 12 January—or at least his own part in it.
'I spoke to them at some length,' he records, and his speech takes up
some four pages of his diary, as against the two lines accorded to the
reasoning of the delegates. 'I persevered in my own course—firmly
believing it to be right ... They hoped for Concession and I made
NONE,' he wrote at the end of his entry for that day.

Sopwith's position was impregnable, provided he remained firm.

T. W. Beaumont died just before the strike, and his successor was a
minor. No interference was therefore possible from that quarter. The
other Beaumont mining districts did not take part in the strike or,
apparently, aid the strikers in any way. Sopwith behaved in a manner
highly infuriating to the miners, spending a great deal of his time away
from Allenheads. He was there for the annual pays on 24 January
and he notes 'that a Five Pound note given in excess of what was due
to a partnership of men, now on strike, was immediately returned to
me'. After that he was away for much of the next two months. In
April there was another attempt by the strikers to open negotiations,
but

Mr. Sopwith, after one of those fulsome orations, at which he is such an
adept, and in which he equally complimented his own liberality, benevolence,
disinterestedness, and our honesty and peaceableness told them that he had
left the entire management of the matter in the hands of Mr. William Curry,
and, suiting the action to the latter words, mounted and rode away, and we
have not seen him since.[25]

The strike was not crippling the Beaumont concern; all the other
mines were working, and the strikers were forced gradually towards
capitulation. The great majority of the men knew they could go back
to work if they were prepared to accept Sopwith's conditions, and the
miners in other districts were already working under these conditions.
They had, of course, to abandon their leaders, some of whom had been
discharged before the strike began, and who had no hope, even had they
the desire, to be employed again by Sopwith. However, these leaders
were very influential. Even Sopwith admitted in his diary that one man,
Joseph Heslop, 'had considerable natural talents'. Their dominion over
most of the men was complete. Dickinson, in his history of Allendale,
says that 'men were afraid to be seen speaking to anyone who had
remained at work, lest it should be thought that they were "favouring
the masters".' He quotes at length from a speech (probably by Joseph
Heslop himself) given at a meeting of the strikers at Swinhope Primitive Methodist chapel on 21 March.[26] Most of it is worth quoting here
to show how strong was the feeling against 'blacklegs' (the term was
noted by Sopwith as being used to describe men who had gone to work).
Something of the passion of the speaker does come over, and his power
over the minds of his hearers may well be imagined.

In the first place I wish you all to know that you all have liberty to go to
work when you think proper. You have it in your power to either keep your

old masters or to have a whole set of new ones. This I can assure you from the best authority. I have a letter in my pocket which I am not prepared at present to let you all see or else you would take courage anew and cry with one voice 'the day is ours' (putting his hand into a pocket like pulling out a letter but did not). Now lads, my orders out of this pulpit to this meeting is that every man has liberty to begin work, but it is my hope and earnest prayer that if any man do begin work in connection with them that has begun, that you will have the goodness to pass by them and their wives and families without speaking to them, to have no connexion or communication with them. If they be sick do not visit them; if they are in need of a doctor do not seek them one; if they die do not bury them; if they are fastened underground in the mines do not assist in seeking them out but let them die, or be killed in the dark, and go from darkness to darkness into the fangs of the devil, to be kept by him without remorse in the fire of Hell for ever and ever. You are all to torment them while on earth, and when they die may the devil torment them to all eternity. Let them be like Cain, deserted by God and forsaken of men. Let them be like Judas, only fit for taking their own lives if none of you do it for them . . . and if they emigrate to Australia or America, if any of you should be there, be sure to treat them in the same way; for I can tell you for one that if I had a houseful of bread and every other necessary of life to take and to spare I would not give one of them a mouthful to save their lives if I saw them dying of want in scores and I hope you will follow my example.

But the leaders really had nothing to offer the miners. Sopwith was not prepared to negotiate; he wanted only capitulation. It must have been this feeling of impotence that prompted the series of letters to the *Newcastle Guardian*. The extracts already quoted show how well they were written, with considerable argumentative skill, and cutting irony. Sopwith himself said in his diary that 'they were drawn up with no common ability and it was shrewdly remarked by some that they were much more like the compositions of a barrister than of working Miners' (he does not elaborate on this suggestion). The later ones consisted almost entirely of attacks on Sopwith. As well as tyranny he was accused (mistakenly) of knowing practically nothing about mining. Some of the remarks, however, have more than a grain of truth in them: 'The leading mental quality which has displayed itself during his whole career among us is vanity, and to this, is doubtless referable most of the evils to which we have been subjected.'

Sopwith's diary shows how well some of these thrusts found their mark. Although prepared to be ruthless in pursuit of what he believed was right, he was not a cruel or a vindictive man. The letters did in fact do him some harm. 'They contained specious arguments which

went a long way in One very influential quarter to produce an unfavourable impression as to the discretion of my line of conduct in the
Management of the Mines.' This was presumably a reference to the
new owner, W. B. Beaumont, one of whose first acts on attaining
his majority in the following year was to dismiss Sopwith (reinstating
him a few months later, however). To this extent the miners' press
publicity won them a victory. But at the time, the attacks had no
visible effect. In April some thirty men were brought in from Alston
to join the few who had already withdrawn from the strike. On 3 May
1849 Sopwith 'observed several groups of Miners on the road near the
Office. I went out and spoke to some of them. They said they had
come to take their bargains!! and this I said they could do—the several
contracts having been all arranged for letting on the 1st of January last.
And thus—after a period of Four Months and three days—the
STRIKE came to an end and the several workmen who could be employed resumed their work.'

About a hundred of the strikers were not re-employed, including all
the leaders, whose letters continued to be published in the *Newcastle
Guardian* until 7 July. Most of them emigrated before the end of the
year. 'While America and Australia spread their bosoms to welcome
the oppressed, we are ready to brave the perils of the Atlantic or Pacific,'
they say in one of the letters. Dickinson recounts that 'about sixty
persons, men, women and children' left East Allendale on one day,
17 May 1849, 'to seek a subsistence on the banks of the Illinois'.

So the strike ended, with the victory for Sopwith and the enforced
departure of all the 'more active and artful persons who had fermented
the mischief'. Sopwith's only defeat was that most of the miners
refused to sign a document admitting their entire responsibility for the
strike, circulated later in 1849. He did not press the point.

The ostensible cause was Sopwith's determination to enforce regular
hours on men accustomed to work if or when they chose. A deeper
reason was the clash of personalities between a strong, determined
chief agent and a small number of equally determined miners' leaders.
If either side had been prepared to yield a little ground the strike might
have been avoided—there was, after all, no trouble in the other
Beaumont mines, where similar regular hours were being enforced. As
it was, both sides almost welcomed the conflict. But the scales were
heavily loaded on the management's side. Work continued in the other
areas, so Sopwith could afford to wait. If the Allenheads men did
attempt to win the support of their compatriots in Weardale and at

Coalcleugh, no record of it has survived. Without such support, they were doomed. When money ran out completely, the strikers had no choice but to capitulate or emigrate. All the leaders took the latter course, and Sopwith was left with a servile population which he dominated (benevolently) for the next twenty-two years.

The absence of trade unionism

This unfortunately incomplete record of industrial disturbances in the lead mining area has shown that each dispute was isolated, both geographically and historically. There is not a single example of miners from more than one concern, or even more than one district, combining to fight the owners. Similarly, there was no continuity between one dispute and the next. Each strike was run by a 'committee' (to quote the mayor of Newcastle in his report of 1819 to the Home Secretary) but the organisation never survived the strike. The Webbs defined a trade union as 'a continuous association of wage earners for the purpose of maintaining or improving the conditions of their employment'.[27] By this definition, no trade unions ever existed in the lead mining region—only temporary combinations, geographically limited, to gain a specific end.

The causes and the objects of the strikes were fairly simple and specific, if the 1849 Allenheads case is excepted. They were normally caused by severe economic depressions which lowered earnings and often raised food prices. This is certainly true of the strikes in Weardale between 1790 and 1820, which occurred after rather than during depressions, at a time when the miners' bargaining position was stronger. It was pointless to strike at a time when such action would serve only to save the employers money. The Teesdale strikers of 1872 probably had similar aims. The Weardale strikes in the late 1870's and early 1880's were rather different, being desperate expressions of protest against the hopeless state of the lead trade at that juncture. The strikers normally wanted higher subsistence and/or bargain prices; sometimes they got both, at other times they got neither.

None of the strikes was revolutionary by nature. There was no Luddism nor machine breaking, and no connection with political issues, although national economic conditions obviously did affect the timing. But the movement for parliamentary reform caused no disturbance in the lead dales; there was no hint of Chartism amongst the miners.

K

There is, however, just one hint of the existence of some potentially militant and trade-union minded men in the mining region. In chapter IX it will be seen how the lead miners flocked to the north-eastern coalfield in 1831–32 to break the great coal strike. 1844 saw a similar strike. In the *Northern Star* of 13 April 1844 there is a report of a meeting of striking colliers at which 'A letter was . . . read by the Chairman from the lead mining districts, requesting information on the subject of the Union and pledging themselves that whatever arrangements the coal miners may make for bettering their condition, they will give them their co-operation, and take care that no lead miners will come and take their places.' It would be interesting to know from whom this letter came, but there is no further information—and no union of lead miners was ever formed.

Their failure to create permanent trade unions to advance their interests was probably the effect of a number of inter-related causes. In the north-eastern coalfield it was the pits which contained the greatest number of migrants that were the most militant. The lead mining areas were characterised by a stable population and a low proportion of immigrants. A very pressing issue was needed to bring the men out in revolt against all-powerful owners who controlled the only form of employment in the area. Another factor was the nature of the relationship between the owners and the miners. The Webbs discovered that 'Only in those industries in which the worker has ceased to be concerned in the profits of buying and selling . . . can effective and stable trade organisations be established.'[28] Even in the last days of the bargain system the lead miners were not orthodox wage-earners. They were selling, not their labour, but its proceeds. Strikes were a last resort.

NOTES

[1] B/B 46, 16 January 1760.
[2] Adm. 66–96, 24 February 1775.
[3] B/B 50, 17 December 1795.
[4] B/B 50, 29 January 1797.
[5] L.L.C. 16a, 13 July 1811.
[6] 1864 report (vol. 2), p. 377.
[7] Quoted by Raistrick, 1938, p. 65.
[8] See Lee, *Weardale memories and traditions*, 1950, for interesting but unreliable stories of this strike.
[9] B/B 50, 27 December 1794.

[10] B/B 50, 23 November 1795.

[11] B/B 50, 17 December 1795.

[12] B/B 50, 2 April 1795.

[13] B/B 50, 17 January 1797.

[14] B/B 50, 17 December 1799.

[15] B/B 51, 30 December 1816.

[16] B/B 51, 25 September 1818.

[17] B/B 51, 2 October 1818.

[18] Printed in A. Aspinall, *Early English trade unions*, 1949, p. 305.

[19] B/B 51, 9 October 1818.

[20] B/B 51, 15 October 1818.

[21] Printed in Aspinall, p. 310.

[22] B/B 51, 24 October 1818.

[23] B/B 51, 26 October 1818.

[24] Striker's letter to *Newcastle Guardian*, 20 March 1849.

[25] Letter to *Newcastle Guardian*, 12 May 1849.

[26] Dickinson, 1903, pp. 48–9. He says that his source 'is a copy, given to the present writer, of a document which was itself stated to be a copy of a report given to the mining agents after the meeting by one who was present'.

[27] S. & B. Webb, *Trade Unionism*, 1902, p. 1.

[28] Ibid, p. 35.

VII

The pattern of settlement

The lead miners' dwellings were situated as close to the mines as geography allowed. In consequence the area of settlement in the region extended to a higher altitude than anywhere else in Britain. Despite the altitude and the climate, many of the miners combined their occupations with the ownership or lease of smallholdings, and part-time farming was an important side of the miners' way of life.

The most significant factor in the siting of settlements was of course the location of the mines themselves. The entrances were high in the Pennines, normally at heights ranging from 1,200 to 2,500 ft. The miners' homes were as near as was climatically and topographically possible. In the Cheviot dales, settlements rarely rose above the 700-ft contour.[1] In the lead mining region several villages were located above 1,000 ft, and scattered settlements were at greater altitudes. The long established villages were high, but not as high as those which grew up mainly in the nineteenth century. Of these older villages, Alston is at 900 ft, Stanhope at 700 ft and Middleton in Teesdale at 750 ft. In the nineteenth century, after the enclosure Acts had made more land available, and when the pressure of an increasing population was making the use of every square yard of inhabitable land necessary, new villages—which were in fact often centred on an existing group of offices with some cottages around them—were placed much higher. Allenheads is situated at 1,327 ft, Nenthead at 1,411 ft and Coalcleugh at 1,821 ft. (There were few houses actually at Coalcleugh; Carr Shield, two miles down the dale, is at 1,312 ft.) Dispersed dwellings, in the nineteenth century, were higher still. There was a lead miner's house in Teesdale at just under 2,000 ft, a remarkable altitude by British standards.

Settlements at these heights required shelter, and the dwellings were sited in the valleys or on the valley slopes of the main rivers and their tributaries—long strips of cultivated land within a moorland waste. Weardale was described in 1842 as having 'houses . . . distributed over it on both sides of the river like a continuous scattered village'.[2] These strips of dispersed habitations grew narrower as the heads of the dales were approached and gradually disappeared altogether. Above Cauldron Snout the Tees flows from its source some fifteen miles away through a vast plateau, without any shelter. On this exposed plain no habitations were built, in spite of the mines that were located there. Over the whole mining area the high land, between the settlement strips in the shelter of the river valleys, was left completely uninhabited.

The greater number of lead miners lived in scattered cottages, not gathered in villages as in the Durham coalfields. The maps of the Allenheads region (plates v and vi) show the large number of individual dwellings located along the sides of the East Allen and its tributaries. Villages and hamlets functioned chiefly as service centres for the dispersed dwellings. A traveller in 1859 stated that 'Allenheads is to be regarded rather as the nucleus, containing a few shops and an inn, of the houses scattered over this part of the dale, than as a village.'[3] In Allendale and Weardale the villages grew in haphazard fashion, developing from long established parish centres as at Allendale Town and Stanhope, or around the offices of the mining company, as at Allenheads.

The village of Nenthead was created by the London Lead Company in the eighteenth century to serve as the centre of its operations on Alston Moor. It remained little more than a group of mining offices and a smelt mill, with a few houses and a school, until the nineteenth century, when it was systematically developed as a residential area for the company's workmen. The company also transformed the older villages of Garrigill and Middleton in Teesdale by rebuilding older dwellings and constructing new ones.[4] Langley mill (of which more shortly) was a planned settlement similar to Nenthead but on a smaller scale, built by the Greenwich Hospital. The older villages were inhabited chiefly by tradesmen and those engaged in 'service' occupations. The 1851 census MS schedules show that in Allendale Town there were about sixty families in this sort of occupation as against only about thirty wholly lead mining families. There were ten tailors, nine grocers, six shoemakers, three drapers, and various individuals in other trades.

By the late eighteenth century there was an acute shortage of suitable land for the expanding population of the region. Plates v and vi show the area around Allenheads about 1800 and in 1861. Comparison of the two shows the great increase in settlements in the sixty years between them. There were many more scattered dwellings at the later period and far more houses in the villages of Allenheads and Dirtpot. Enclosure of the commons made this expansion possible. In the Park and Forest quarters (the Western area) of Stanhope parish, the demand for land resulted in much fragmentation of copyhold holdings before the Weardale Enclosure Act of 1815. In 1762 there were 105 proprietors: in 1803, 185; and in 1856, 256. Between 1762 and 1803 there had been no legal expansion of the area available for occupation so the increase represents mainly a greater dividing up of the same amount of enclosed land. (There were probably also a number of illegal encroachments on the commons.)

After 1815 much more land was officially available, and a quarter of the dwellings still standing in Stanhope parish were built between 1815 and 1840.[5] In 1852 Sopwith noted in his diary a selection of the normal and frequent problems he had to deal with. One was finding houses for those in need, 'for in these Mountain wilds, inhabited only by Miners and those dependent on Mines population presses closely on habitations'.[6] In the nineteenth century the London Lead Company solved the problem by building cottages, and by letting land for its employees to build them on in the villages of Nenthead, Garrigill and Middleton in Teesdale. The 1851 census MS schedules show that by this time the tradesmen of Nenthead, unlike those of Allendale Town, were greatly outnumbered by the lead miners living in the village. After 1860, when the industry lapsed into its final decline, the problem disappeared with the enforced migration of many families.

Housing

The actual homes of the miners were all built of stone—mainly the easily workable sandstone available from deposits all over the lead dales. Bishop Pococke found that in the Stanhope area in 1760 'they thatch their houses with a thick coat of heath, and make the roofs steep that the melted snow may not soak into the thatch, and they lay loads across the top of it to keep out the water'.[7] The 1842 report said that the houses in the Stanhope area nearly all had slate (or 'flag') roofs.[8] There had been a definite improvement in building materials and methods during the intervening years. Few domestic buildings still extant in the lead dales appear to date from a period much earlier than the eighteenth century. This impression is confirmed by a detailed architectural survey of 'small house' architecture carried out in the nearby Eden valley.[9] An examination of the homes of the Cumberland 'statesmen', a class of small farmers bearing some resemblance to the lead miners with smallholdings, has shown that not a single house has survived from before 1660. It is probable that in both areas the houses of the lower classes before this date were too insubstantial to survive.

The isolated dwellings occupied by the majority of lead miners were designed to serve as small farmhouses as well as homes. The usual pattern, as witnessed by surviving examples (most of which have been heavily rebuilt and altered), and from contemporary descriptions, consisted of a long two-storied building. The ground floor was occupied at one end by a large kitchen-cum-living room, and at the other by a cattle shed. Above the kitchen would be one or two bedrooms, and above the cattle shed a hay loft.[10] Sometimes two or more dwelling houses were set in a row, each with its associated outbuildings. A plentiful supply of lime made whitewash cheap, and it was used liberally. In Teesdale, according to Sopwith in 1833, the whitewashing 'partly redeems the poverty of their aspect, and this operation is said to be always performed, with becoming loyalty, on the approach of the most noble Duke [of Cleveland] to the moors in the shooting season'.[11]

When the Greenwich Hospital built its mill at Langley in the 1760's a great deal of discussion took place about the design of the workmen's cottages. Three types of house were considered in 1769. The prices given here are for pairs of semi-detached houses.

The first 50 by 10 feet Outside and 13 feet high is upon an Idea that the Persons living in them are to have Cows and the whole Apartment for each Family in this way will be:—
One room on the Ground Floor 16 by 15 feet.
A Milk house or Closet on ditto 4 by 6 feet.
A Byer or 2 stands for Cows on ditto 11 by 6 feet.
One Room above to communicate with the lower room by a step ladder 22 by 15 feet. £152. 13. 8.
The second have no Convenience for Cattle, but have a Stair Case Common to each House—The Apartments as follows for each Family:—
One Room on the Ground Floor 16 by 15 feet.
One room above 16 by 15 feet. £146. 0. 10.
There will be a place under the Stair Case and another above it and it is proposed that the Familys living in two Cottages shall each take one of these Conveniences. You'll observe that in this way the two houses lye very much open to each other, which is an objection certainly and you'll see the only reason of doing the thing in this manner is to save expenses. But still to save

more and surely more agreeable to mere Cottagers who like to live by themselves is the third.—The Apartments for each Family will be:—

One Room on the Ground Floor 18 by 15 feet.

One room above to communicate with the Ground Floor by a step ladder 18 by 15 feet.   £133. 9. 10.

In every case the upper Rooms are to be Ceiled.[12]

It was decided to build houses of the last two types although, as we shall see, it was later found necessary to add the farm outhouses after all.

There is more evidence about the design of the nineteenth-century cottages in the lead mining areas, particularly those in the London Lead Company villages. In 1864 Dr Peacock looked over examples of the cottages in Nenthead.[13] He found that the best houses, which were occupied by the smelters, had adequate drainage, a coal cellar, dustbin and privy. But most of them had only two rooms, although the company had just finished building some new houses with two rooms on each floor. The houses in the village that had been built by the workmen themselves on company land were 'defective, and some of them are most objectionable'. There were never more than two rooms, sometimes very small and low. Some still had thatched roofs, without proper ceiling for the upper room. Sometimes there were no privies. References in the company's minute books show that the court was conscious of bad housing conditions and was attempting to improve them.[14] This was more than most of the other mining concerns were prepared to do, so that conditions at Nenthead in 1864 were probably superior to those elsewhere in the lead dales.

The effects of overcrowding and bad sanitation were somewhat mitigated by the fact that every settlement was surrounded by a wide expanse of moorland. The only form of air pollution was caused by fumes from the smelt mills, which were highly poisonous and destructive of animals and vegetation, both as a vapour and as a sublimate. After the establishment of Langley mill, many complaints were received that the fumes were killing off the cattle. Two cottages had to be built very close to the mill to guard it against theft. 'The situations will undoubtedly be extremely unwholesome, but by letting the Occupiers have them rent free, and allowing them Coals and Peats for their Fire, it is hoped that some of the Workmen may be induced to inhabit them.'[15] This particular problem was largely removed at the end of the eighteenth century by the construction of long, horizontal chimneys taking the fumes far away from inhabited areas. Even in 1842, however, the wind sometimes carried them back from the mouth of the chimney to the village of Nenthead, which was most 'disagreeable' to the inhabitants.[16]

Judged by more modern standards, housing conditions were not very good. Nineteenth-century observers, however, found them better than in most agricultural districts of England, and considerably better than working-class conditions in towns. Sir John Walsham's series of reports to the Sanitary Inquiry Commission of 1842[17] show that most agricultural labourers in the north lived in single-room cottages. Around Alnwick many consisted of nothing but 'a rough room of lime and stone, covered with thatch, and with nothing but an earthern floor, except a flag-stone or two near the hearth'. Around Hexham the labourers' cottages 'rarely exceed one room, with such division in it only as can be effected by the arrangement of the furniture of the occupant'. The 1864 Hunter report on rural housing concluded, 'The majority of Northumbrian and Durham peasants, whether rich or poor, hind or collier, live in but one room, day and night, with all the family'. In his visits to 224 rural houses in these two counties (he did not reach the lead mining districts), Hunter had not found a single example with more than one bedroom.[18]

Witnesses before both the 1842 and 1864 commissions affirmed that housing standards in the lead mining area were superior to those in the surrounding agricultural districts. Severe overcrowding was unusual. The average number of persons living in the 169 dwelling houses in Nenthead in 1851 was five; only twenty were inhabited by eight or more people, and only six by ten or more. The largest number of people in any one house was twelve. The census schedules show similar figures in Allendale. Sopwith, however, told the 1864 commission that some overcrowding was caused by the households nearer the mines taking lodgers.[19] The houses of the lead miners seem to have been of considerably better quality than those of the colliers in the average pit village. Hunter said, 'The lodging which is obtained by the pitmen . . . of Northumberland and Durham is perhaps on the whole the worst and dearest of which any large specimens can be found in England.' He based this judgment on 'the high number of men found in one room, in the smallness of the ground plot on which a great number of houses are thrust, the want of water, the absence of privies, and the frequent placing of one house on top of another'.[20] The lead miners' dwellings were mostly dispersed, on a mountainous terrain with plentiful supplies of water, and where the inconveniences of poor sanitation were not so noticeable as in the badly drained pit villages.

The high wages of the pitmen, however, allowed them to furnish their houses in a style far beyond the pockets of the lead miners.[21]

The tenure by which the lead miners held their houses and land was of three kinds: freehold, copyhold and leasehold. On earning a substantial sum of money by a good pay, or by the steady accumulation of small sums, many lead miners bought or leased a small patch of land on which to build their own homes. But when times were bad these homesteads frequently had to be mortgaged and possibly sold. In 1834 most owner-occupied small properties in Stanhope parish were mortgaged.[22] A relatively small number of miners owned their own houses and land, and it would appear from the evidence about enclosure (see below) that in the nineteenth century their numbers decreased relative to those who leased their land. Most landlords operated the system in use by the London Lead Company at Middleton in Teesdale. Miners were permitted to build houses at their own expense on company land, and were then subsequently charged rents for them.[23]

The great landlords in the lead mining area exercised even more power over their tenants than was usual in the eighteenth and nineteenth centuries. The biggest landowner in each of the mining parishes was the lord of the manor concerned—in Allendale the Blackett/Beaumonts, in Alston the Greenwich Hospital, in Middleton in Teesdale the Duke of Cleveland, and in Stanhope the Bishop of Durham. In addition to the land, they owned the mineral rights, and this fact allowed them to search for lead more or less regardless of any inconvenience to the surface occupiers. The enormous acres of common land in these parishes belonged, in law, to the lords of the manors concerned, giving them the greatest influence on the vital question of enclosure.

Copyhold tenure was more usual than freehold, and holders of land by this tenure still had to fulfil certain feudal obligations to the lord of the manor. In Weardale and Teesdale the landowners were not directly interested in the mines, contenting themselves with collecting royalties from those who did the actual mining. In Allendale the Blackett/Beaumonts extracted the lead themselves, and so their tenants were also their employees. The Greenwich Hospital's employees at Langley mill were also the Hospital's tenants. On Alston Moor, and to a lesser extent in Teesdale and Weardale, the landowners leased much surface land to the mining companies, and so the latter were frequently their employees' landlords.

Rents were not low. In 1794 it was said that the Durham lead mines 'do more than double the rent of all the small farms contiguous thereto, as the miners take these farms at extravagant rents, for the convenience of keeping two or three cows and a galloway'.[24] Eden, in 1797, confirmed that rents were very high compared with those in the areas around.[25] But they were often lowered in times of severe depression. In the early 1830's the price of land had dropped with the price of lead, and many landlords had given up trying to collect their rents.[26] They would have gained nothing by ejecting the miners from their homes, as there would have been no tenants to replace them any more capable of paying the rent. It was this lack of competition for their houses and land from local farmers that gave the lead miners their only source of strength in dealing with the landlords.

Smallholdings

A Weardale writer, describing the normal routine and life of lead miners in 1840, said 'Their work is among lead ore, sealing pastures, waiting upon and feeding cattle, mowing, winning and stacking hay, and carting fuel against the winter season.'[27] In other words, most lead miners combined mine work with agricultural labour on their tiny upland farms. Many contemporary writers commented on the lead miners' desire for land, which led them to pay high rents for ground that elsewhere would have been left as waste. Their farms were very small, contrasting with the larger farms further down the dales. 'The great feature in Teesdale is the small size of the farms . . . As one gets higher up the dale the holdings diminish in size; a portion of the family or lodgers work at the lead mines.'[28] In 1834 the Poor Law Commission was told that in Weardale and Teesdale the 'miners' farms consist of about three acres of meadow, three or four of upland pasture, with a house and offices necessary for accommodating two cows, or a cow and a galloway, and are such as the miners have spare time to attend to without trenching on their ordinary labour'.[29] The average acreage of each farm was probably rather smaller. There appear to have been no noticeable differences in the farming settlement from dale to dale. In the eighteenth century, before parliamentary enclosure took place, the pattern was much the same, with more grazing on the open moor instead of enclosed upland pasture.

Statistical detail as to the numbers and proportion of lead miners with smallholdings and the average size of holdings is very difficult to assemble. The enclosure awards for the different parishes show the number of people holding land by free or copyhold; they do not show those who held land by leasehold, as large numbers of the lead miners did. The tithe appropriation surveys of the 1840's are also disappointing as evidence, since they do not analyse the tenants holding small amounts of land but merely note the landlord and state that the land was let to one named person 'and others'. The best source, analysing the leaseholders of carefully defined and mapped areas, is an estate survey and rental of the Beaumont estates in Allendale carried out under the direction of Thomas Sopwith in 1861. It consists of a specially surveyed and printed map, showing all structures and enclosures, together with a key listing all the tenants, giving the precise size of their holdings (analysed into meadow and pasture) and stating the number of stints, or grazing rights, they had on the unenclosed commons. Of the 44 houses in the Coalcleugh area, 25 had enclosed meadow or pasture land in connection with the house, although not necessarily immediately attached to it. In the Allenheads area the figure was 81 out of 123. Thus at Coalcleugh 56·8 per cent of the houses had smallholdings, and at Allenheads 65·8 per cent.

As regards the acreage of the smallholdings, most of those at Allenheads and Coalcleugh seem to have been between one and ten acres in size, with rather more under five acres than over. It is noticeable that the larger holdings consisted of much more pasture than meadow, while the smaller were often only meadow. The allocation of stints bore no apparent relation to size of holding, although no house without any land at all had any stints.

It would be wrong to draw too general conclusions from these figures, relating to two small areas only, both owned by the same landlord. Undoubtedly the position in the eighteenth century, before parliamentary enclosure, would have been very different. But evidence from the various parliamentary reports suggests that the proportions of miners with smallholdings, and of the amount of acreage held, were much the same over all the lead dales in the nineteenth century. Unfortunately most of the statements on this subject made by witnesses in 1842 and 1864 were vague, such as 'a great many of them have little farms'.[30] In 1857 the chief agents of both the Beaumont concern and the London Lead Company were closely questioned by members of the Select Committee on the Rating of Mines as to the relative numbers of people engaged in mining and in agriculture in the lead mining region. Sopwith replied, 'There are very few purely agricultural labourers; most of the farms are occupied by miners or smelters, and people connected with those works.'[31]

Bainbridge, the chief agent of the London Lead Company, replied similarly that nearly all the farms in the upper dales were in the possession of lead miners, and further pointed out that 'the population is so mixed up, the farming with the mining population, that they are almost all as one; it is scarcely possible to go into a family occupying half an acre in Alston Moor or Teesdale, without finding that one or more members of the family are workmen employed in mines.' On Alston Moor, where the Greenwich Hospital was ready to let land on long leases, 'many of these parties who are in possession of estates . . . are fortunate miners; miners who in past years have speculated and have been fortunate, and who have bought up little properties around them.'[32]

The miners were able to work their small farms in addition to their normal occupation because their working day was short. In the eighteenth century a six-hour day was common, and even when an eight-hour day became usual in the nineteenth century the hours were sufficiently short by contemporary standards for plenty of time to be spared for agricultural work. Many of those miners who lived away from home in accommodation near the mine during the week arranged their shifts so as to finish work by Friday morning, to enable them to have a long weekend at home. In any case, no adult miners normally worked on Saturday or Sunday in either the eighteenth or the nineteenth century.

The men were allowed to take a few days' holiday at the time of the hay harvest. In the mid-nineteenth century 'five weeks about the time of the hay harvest' were officially given as a holiday to all children attending the Beaumont schools. Since few of the miners' wives had any occupation even in the eighteenth century, their labour was available for the running of the farm. Older men, whose health had been ruined by the respiratory diseases endemic in their work, often retired to their smallholdings, leaving their sons to be the wage-earners.

The smallholdings meant to the lead miners not only the supply of a certain amount of food but also a welcome contrast to their labours underground or in a smelting mill. They were moreover a concrete sign of success, and a buttress against bad times:

Occupations of this sort give the men something more than grooves, ore, lead, washing and smelting, to talk and think about: in short, they become small farmers as well as miners, and so far as my observations have gone, interest themselves wonderfully in the practice of agriculture; so that in the intelligent management of their meadows and stock they compare favourably with professional farmers in some parts of the country.[33]

If the smallholding tradition was agreeable to the miners, it was also advantageous to the landlords and owners. The men's desire for smallholdings made it possible for the landowners to let land which in normal circumstances would have been left waste. For the mine owners, the existence of smallholdings attached the labour force to the district more strongly than anything else. In his diary for 1866 Sopwith recorded his disquiet at the policies of Mr Beaumont's land agent in Allendale; he was raising rents steeply, and objecting to reasonable requests for repair and maintenance grants. Sopwith wrote, 'If by raising rents and being chary of repairs the revenue of the land is increased, yet I much fear that in a still greater degree are the permanent mining emoluments endangered.' The bulk of the estate should be considered as merely

a useful adjunct to the Mines. I think the small holdings of Cottages and of grass for a cow or even a small garden have had an important influence in making the people attached to the place—willing to work in it and unwilling to leave it. As regards the tenants I feel satisfied that many of them *place a far higher value on a cheap holding* of a farm or cottage which is dear to their associations *than they would on the receipt of cash equal or even double or treble the value of their holdings*.[34]

The landlords and mine owners encouraged the miners to own a little plot. Such a policy was aided by the fact that there were few strong economic or agricultural arguments in favour of using the land in any other way. The Greenwich Hospital carefully considered the arguments for smallholdings when Langley mill was established. In April 1768, before the mill had been built, the receivers notified the board of the Hospital that 'as the Mill is at a distance from any village, it will not only be necessary to build an House for the Agent, but also some Cottages for the Smelters'.[35] By January 1769 the time for detailed planning of the designs for the cottages had been reached, and the three alternatives already quoted were submitted to the board. In the covering letter the receivers said that 'the Men are some of them very desirous to have little pieces of Ground to enable them to keep cows'. It was not advocated, however, that the cottages should initially be designed for this purpose, as

at this Rate every Workman must have a Farm and therefore it seems to me that we should only furnish conveniences such as Smelters etc. are content with where they have no Land. . . . It may, and will it is most probable, be necessary to build more Conveniences of this kind, and as such it is the more proper for us to be Careful not to introduce a thing by which we may have a difficulty, and even supposing that it should hereafter be found an advisable thing to accomodate the Workmen with pieces of Land, Housing may be built for their Cattle without any material loss.[36]

The houses were therefore built without any farm outhouses attached.

Once the cottages had been built and were occupied, there was immediate pressure from the smelters for some sort of farm accommodation to be provided. Some of them apparently already kept cows on the commons, for they complained that 'among those that keep cows now, having everything to buy, it comes so expensive to them, that they cannot bear it; and being far from a Town or Neighbourhood they are under difficulties in making provision for themselves'.[37] The smelters' skilled labour gave them considerable bargaining power, and three years later the receivers gave in to this pressure. Land was to be taken from a neighbouring tenant farmer of the Hospital, and let to the cottages at the rate of 6s 6d an acre. 'We took this trouble merely for their accomodation, as we understood how very Desirous they were to have Allotments of Grounds in their own hands.'[38] The chief agent of the mill was asked, before any action was taken to secure the land, to ensure that 'it will be of real advantage to the People: for we shall not chuse after so much trouble as this will give us to have them complaining of hard Bargains; 'tis far better they should complain for want of land, than of having land by which they are loosing'.

The smelters obviously agreed; byres were added to some cottages, and an old cottage was wholly converted into sheds for the farms. The fields were enclosed (see plate VII), and the cottagers were to pay for the cost of the necessary surveying, and the work of enclosure, plus 5 per cent interest, all to be added to the initial rents. As there were not enough fields to go round, it was arranged that the refiners would have first choice, and lots were to be cast as to who should have what land. The land was divided into small enclosures, but two upland pastures were left for common use. Ploughing was not permitted. Six months' notice to quit was required on either side.[39]

The housing and farming policies of the other mining bodies are less well documented. In the eighteenth century the London Lead Company followed a policy of *laisser faire*, allowing its employees to find or construct their own habitations.[40] Some time after the company took over leases on Alston Moor, in the late 1730's, a smelt mill was

constructed on the river Nent, at what was to become the village of
Nenthead. Little is known of its eighteenth-century history; probably
it started and developed as a settlement in much the same way as
Langley, for lists of claimants for land in this area in the enclosure
records of 1815 suggest only a few dozen inhabitants at the time.[41] By
the end of the eighteenth century, however, the company was actively
assisting its employees on Alston Moor to obtain houses and smallholdings, itself often buying or long-leasing land for this purpose.
Around Garrigill, cottages were built for renting, with up to six acres
of enclosed land attached, together with grazing rights on upland
pasture.

In the nineteenth century the company changed its policy to one of
deliberately developing village centres. After 1820 Nenthead was
systematically rebuilt, as also was the older village of Middleton, in the
newly acquired Teesdale area. Not just houses were built in both, but
also such community centres as schools, chapels and municipal buildings. In 1851 Nenthead had a total of 872 inhabitants. By the middle
of the century the company was providing gardens with every cottage
built, as a substitute for farming land. Cottages built by the miners
themselves, within the company villages, and some of the earlier
company cottages, did not always have gardens.[42]

The Blackett/Beaumonts never concerned themselves very much
with the houses and small estates of their work people, except as a
source of revenue. In the first part of the eighteenth century their
mining agents were also their land agents. By the end of the century
a separate land agent at Hexham was responsible for the estates. This
splitting of authority was much regretted by Sopwith when he became
the chief mining agent, but he could do nothing about it. He would
have liked planned expansion of the settlement areas, along the same
lines as the London Lead Company's policy. As it was, expansion in
East and West Allendale and Weardale remained as haphazard in the
nineteenth as it had been in the previous century.

Virtually nothing is known about the policies of the smaller mining
groups. In the eighteenth and early nineteenth centuries, when there
were many tiny groups of adventurers working on Alston Moor, 'the
first profits of their success are laid out in the acquisition of a Cottage
and piece of Land within the Manor, which secures to them a comfortable Home, and supplies to them the Means of conducting their
speculations with increased confidence.'[43] Later in the nineteenth
century many of them had land available to rent to their workmen.

In 1864 the Rodderup Fell Company, for example, was letting houses together with a right of pasturage for two cows on the common.[44]

To sum up, then, both the employers and the employees in the lead dales favoured the smallholdings system. The miners liked the real assistance their holdings gave in supplying their families with food: the welcome relief and relaxation from underground and mill work; and the (perhaps illusory) feeling of independence. The landlords and mine owners approved partly because the land was not much use for other purposes and partly because it made their employees more firmly attached to the area. The percentage of miners with smallholdings at any one time cannot be calculated. Probably rather more families had holdings than not. Pressure for the limited amount of suitable land made it impossible for all miners to have them.

Farming

Farming was largely pastoral in nature. The high altitudes of the settlements in the lead dales meant that most of the land was unsuitable for arable farming. A nineteenth-century farming manual described the soil of Alston Moor as 'well known, by those whose lot it has been to cultivate it, to be an ungrateful and unprofitable soil'.[45] There is, however, some good loamy soil along the bottom of the valleys and for some little way up the sides. The line of the outcrop of the great limestone formed the boundary between the cultivated and uncultivable ground. Below the limestone the fields would grow lush grasses, suitable for cutting into hay, but above it the land consisted of heather-covered moors with wide expanses of peat bog. Uncultivated moorland predominated. In Alston parish in 1834, for example, the Poor Law Commission was told that in its 25,000 acres there was 'scarcely any Arable, not much Wood land; between 2000 and 3000 acres of Grass Land of different qualities, and the remainder Pasture and Wastes'.[46] Some corn was grown in the lower parts of the dales, farmers being tempted by the high prices grain commanded in this isolated area. These, however, were all full-time farmers, not part-time miners. After the decline of lead mining and the partial depopulation of the area, corn growing was largely discontinued and the land given over to grass.

The lead miners' small farms were devoted only to grass. Stock was kept for two main purposes—to provide food and to act as transport. The range was limited by the fodder available: poultry and pigs, for example, which being easy to rear would have seemed a natural choice, were in fact rarely kept because all the cereals and potatoes were needed to supply the miners' own diet. Geese, which feed largely on grass, were common, as were bees, which could fend for themselves amongst the heather.

Most miners who had smallholdings kept at least one cow, many two or more, according to the size of their holdings. Lists of tithes in Allendale in the 1760's and 1770's show that nearly every person who paid tithe owned one or more cows.[47] These cows would supply the families with milk, butter and cheese. In summer the cattle were grazed on the open moor or in enclosed rough pasture. In winter they were kept on enclosed land and supplied with hay. Calves were reared in the summer, when milk was most plentiful, or, if born in the winter months, on such makeshift food as oatmeal-gruel and hay tea.[48]

Horses, or rather the strong and hardy Dales ponies, were reared in a similar manner. In winter they were frequently sent down to the farms at the foot of the dales, where, in 1842, farmers would keep them for a rate of 1s or 1s 3d a week.[49] The ponies served the need of the miner and his family for any transport, either as a galloway or for drawing a cart. In the eighteenth century, at least, a horse was almost essential for those living in isolated parts of the dales, 'the Markets for Corn and other Provisions which they have to purchase for the support of their Families being at a great Distance, besides having their Hay and Peats to lead.'[50]

Sheep were kept on some of the larger smallholdings, entries for them in the tithe registers being about a fifth of those for cows. It would also appear that these miners made a little extra money by the sale of lambs and wool. Goats—more hardy than sheep, and also yielding milk—were common all over the lead dales.[51]

The enclosures

In the late eighteenth and early nineteenth century, vast amounts of land in the mining parishes were enclosed by private Acts of Parliament.[52] These Acts were the result of pressure from the large landlords in their desire for the improvement of their estates and more efficient methods of agriculture, not of the wishes of the smallholders. Their effect, however, does not appear to have been entirely contrary to the smallholders' interests, since the amount of land available for settlement was increased without, apparently, any mass expulsion of smallholders to engross estates. Farming efficiency was stimulated, and sufficient common, or stinted pasture, remained to satisfy the inhabitants' needs for grazing and fuel. Against this there appears to have been a gradual decline in the numbers of small free and copyholders, and the transfer of their tenures into leaseholds.

The basic pattern of farming was not changed by enclosure. It was practically impossible to keep animals without some land enclosed. The enclosed meadows were needed to provide hay in the summer, and as winter pasture. The open moors would not support livestock all the year round. Consequently there was much demand for enclosed land by the increasing population of the dales in the late eighteenth century, with no legal method of meeting it other than by the fragmentation of existing smallholdings. After enclosure, the amount of land available to serve as meadow and winter pasture increased, and the number of settlements with it.

The terms of enclosure varied from Act to Act in minor details, but were sufficiently similar in the lead mining parishes to enable them to be discussed as one. In each parish, land 'best situated and most capable of Cultivation and Improvement' was to be set out and allocated to those already possessing rights of common in proportion to the amount of land they owned, whether by free or copyhold. If land had been encroached from the common by a landowner in the past, it was to be regarded as his if he had held it for thirty years without legal challenge (in Alston and Weardale; in Allendale the period was fifty years); if, after examination by the commissioners executing the terms of the Act, it was discovered that the land had been held for a shorter time, it was to be given to the holder as part of his share of the land to be divided provided it was not more than his fair share. The areas left unenclosed were to be stinted pasture; stints, or grazing rights, were given to each landowner in proportion to the amount of his holding. These stints were calculated by the enclosure commissioners on the basis of ensuring that the moors were fully grazed but not over-stocked. Smallholders in Allendale and Weardale could receive *either* land *or* stints if they so wished. In Alston they could receive a monetary settlement instead if they preferred. The commissioners were empowered to lay out routes for roads. All the allotments had to be fenced or walled within a certain period of time. Quarries of stone and lime were left for common use by landowners and their tenants. Peat could still be gathered from the stinted pastures, and the mineral rights of the lords of the manor over each parish were not affected by enclosure.

The immediate physical effects of the enclosure Acts and the awards resulting from them were the building of miles of stone walls, and a great impetus to road construction. The effect on the population was less obvious. In 1811, the London Lead Company's chief agent wrote of Alston Moor, 'the division of the Commons has brought the Country into a state of the most abject poverty'.[53] The subject was as hotly debated by contemporaries as by modern historians, and one man's opinion is not sufficient evidence on which to form a reliable judgment.

In the papers of the commissioners for Allendale[54] there is a letter dated 7 April 1799 from Peter Mulcaster, the Greenwich Hospital agent, who owned land in Allendale, making much the same point as the Lead Company's agent; with it is the draft of the commissioners' reply, attempting to rebut the charge and successfully showing up the probable exaggeration in Mulcaster's claims for the productivity of his land before enclosure. Mulcaster's land was let to four tenants, at a total rent of £38 a year. These four had eleven cows between them, and one horse each, together with a small number of sheep, with 'no other Pasturage for either Cows, Horses, or Sheep except the Common.' He had hoped to receive after the division, together with a small allotment, as many stints as were necessary 'for as much Pasturage as the Farms would winter'.

In fact, he alleged, his allotment was much smaller than he expected, and the stints were too few for this purpose; his land would no longer be able to support as many animals as before enclosure. 'I cannot help thinking myself worse used than any other tho' a complaint from most of them is made.' Another piece of land he owned in Allendale received only three and three-fifths stints (enough for two horses); this was tenanted by 'an Ore Carrier of Langley Mill who has always kept from Eight to Ten Galloways grazed upon the Common in the Summer Season, but now as soon as the Award is settled you see that must be over, he can keep no more Carriers and, I fear, it will be the same with nearly half of the present Ore Carriers'.

The complaint was answered in a letter from one of the commissioners, John Fryer, dated 11 April. He pointed out that Mulcaster's ancient estate contained only 35 acres of poor ground, and questioned whether in fact it could ever winter the stock Mulcaster claimed his tenants possessed. With regard to the size of the allotment Mulcaster had received, 'for your estate of 35 acres we have given you 55 acres of the best and most convenient Common that we could possibly find, and 14 stints upon the undivided part'. With regard to the other estate let to the 'Ore Carrier of Langley Mill, who has always kept from 8 to 10 Carrier Galloways—This Estate contains 13 acres—10 of which is the worst land in Allendale; for this we have allotted you 16 acres of Common, part of which is considerably better than the Inclosed Land.' In a subsequent passage he quotes Mulcaster's letter, saying that the 'Division of Allendale Common will be a very great loss to the small proprietors of Lands there. . . . I know the fact to be exactly contrary —I am convinced there is not a single small Proprietor in Allendale (yourself not excepted) but what will be considerably benefitted by the Division: this is my assertion against yours.'

This exchange illustrates the passions the enclosures gave rise to. Fryer's reply appears convincing; Mulcaster almost certainly exaggerated the amount of stock his land had been capable of supporting. If the figures he gives were correct, his tenants would have lost a considerable amount of grazing on the commons which remained unenclosed, owing to the stintage limitation. In Allendale one stint represented one cow, five sheep or eight lambs. Two stints were needed for each horse. But, as Fryer said, it was highly unlikely that Mulcaster's tenants, particularly the ore carrier, could have wintered the stock he claimed on their holdings.

There is no evidence that the original enclosure Acts forced small free and copyholders to sell their land. In the terms of the Acts, they were offered the choice between land and stints if they so wished, but there was no compulsion. The expense of walling, often alleged as a cause of great hardship to the poorer freeholders, may have caused some difficulties, but the miners had the free time to perform this task themselves, and there were plentiful supplies of stone.

There was one important source of loss to the smallholders; its exact importance is impossible to estimate. This was encroachment upon the commons before enclosure. In Allendale the agent of the lord of the manor (Thomas Richard Beaumont) gave the enclosure commissioners a long list of objections to certain pieces of land being recognised as the property of the occupiers. The list totalled 136 items, the great majority of which were in the hands of smallholders. The encroachments varied from a 'dwelling house in Allendale town' to individually named fields. This list shows how extensively the Allendale commons had been encroached upon before parliamentary enclosure. Presumably the same state of affairs obtained on all the commons in the lead dales.

The extent to which these objections were upheld is unfortunately not indicated in the surviving manuscripts. If the encroachment had taken place more than fifty years previous to the Act, it was admitted; if subsequently, it could be included as part of the free or copyholder's allotment. If, however, the encroacher had no other land, it would have been taken from him completely. It is possible that this form of confiscation was what the London Lead Company's agent was referring to when he alleged hardship on Alston Moor as a result of enclosure.

Most of the enclosed land was allotted to the large landowners in each parish. Plates v and vi show what happened to these large allotments in the nineteenth century: they were broken up among smallholding tenants. The estate survey of 1861 shows that not only were more isolated individual cottages built within the new small-holdings thus created, but also fields away from houses were leased to those living in the village settlements of Allenheads and Coalcleugh.

During the nineteenth century in Allendale there was pressure for further enclosure of the remaining commons by the landowners with large or medium sized estates. Sopwith's diary records many meetings on this subject in the 1850's and 1860's, as well as the increased pressure by the larger landlords to buy up the estates of the smaller ones after 1850. In Allendale parish in 1829 there were 236 stint holders, 212 with fewer than twenty stints, and 179 with fewer than ten. In 1851 there were 205, 149 having fewer than ten stints. By 1856 the total number of stint holders had gone down to less than 180.[55] But in that year W. B. Beaumont resisted the calls for a new enclosure Act. He was against curtailing his mineral rights, even in a minor way, and grouse shooting—an increasingly popular sport of the Victorian gentry—would be hindered by more enclosures. The smaller stint holders were united against them too. They did not want the trouble and expense of walling, and the smaller the area of open moorland became, the smaller also became the area where they could dig peat and quarry building stone.[56] In 1871 a Bill for further enclosure in Allendale was presented to Parliament. Stint holders who owned fewer than thirty stints would have been compelled to sell them to the highest bidder. It was not passed.

Enclosure did result in a reduction in the number of free and copy-holdings. As the land was mostly unsuitable for large-scale farming, this did not mean the wholesale dispossession of smallholders but rather an increase in the number of leaseholders. The 1861 surveys of the Beaumont lands in East and West Allendale and Weardale show that nearly the whole of the upper parts of each dale were let to small-holders. Fragmentary evidence for Teesdale and Alston Moor shows that the same condition existed there.

Improvements in farming methods

Enclosure stimulated good farming practices among the smallholders. By the nineteenth century there were three distinct types of land in the lead dales. First and greatest was the vast area of unenclosed and uncul-tivated moorland, surrounding all the dales in horseshoe-shaped belts. Lower down the hill slopes there was a second belt formed of rough pasture grounds. Then, in the bottoms of the valleys and reaching up the slopes on either side, was meadow land, producing one crop of hay a year and providing some winter pasture. After enclosure, and more particularly after the mid-nineteenth century, many of the small-holders made a concentrated effort to transform their holdings from the second type of land into the third. Fortunately, a writer in the *Royal Agricultural Society Journal* for 1868 gave a detailed account of the methods used in improving lands in 'one of the lead-mining districts of Northumberland' which, from internal evidence, must be Allendale.[57] The account refers to a period fifteen years before it was written.

It described the contrast between the reclaimed and unreclaimed lands, although their soil was basically similar:

The process of reclamation consisted in burning and pulling the heather, paring off and burning the turf, and spreading the ashes thereof on the land. The draining, which was done with stones, only aimed at the removal of springs. . . . A good dressing of lime was applied, as much as from 5 to 10 cart loads to the acre. Where peat-earth or any other good earth could be got conveniently, it was mixed with the lime to make a compost, and cowhouse manure was afterwards used.

Some fields were also covered with a network of trenches three or four feet deep. All this work increased the crops of hay the fields would grow, and enabled the stock of the smallholding to be correspondingly augmented. Some fields had been treated in this way fifty years before (i.e. shortly after the enclosure Act), but the work was still going on in the 1840's.

Similar improvements were carried out in the other dales, particu-larly in upper Teesdale. In 1881 'land so reclaimed originally worth

1/– to 3/– is now worth 20/– to 25/– an acre', and would 'certainly graze three times the stock, and not only so, but the class of animals that can be kept is very superior'. This 'remarkable improvement' had taken place mainly in the previous twenty years. In Weardale the process of reclamation had not gone nearly so far.[58] This raising of standards was matched by such external factors as better roads and new railways. Moreover, by the mid-nineteenth century there were energetic agricultural societies in all the districts, supported with enthusiasm by both employers and miners.

During the final decline of lead mining in the 1870's and 1880's, smallholdings gave the miners some sort of protection against economic hazards. No doubt they made migration a particularly heartbreaking decision. But the smallholdings were essentially a part-time occupation only. The older members of the family might devote all their time to farming, but it produced no monetary income; only a portion—and that not the largest portion—of a lead mining family's food could come from its smallholding. A small farm was no viable alternative to the miner whose main occupation had gone. Unless he were fortunate enough to obtain other employment in the district, he had to emigrate.

> When . . . the depression of the lead trade . . . set in, many of the small farms became rather a burden than otherwise to their occupants, who were compelled to find employment too far away from their homes to enable them to return there each day, and were consequently given up. The result is that one individual may now be found farming a combination of holdings which were formerly occupied by seven or eight different tenants.[59]

This is Allendale at the beginning of the present century, but the same thing happened in all the lead dales, and some of the very high land was completely abandoned.

The pattern of miners' smallholdings, however, greatly influenced the twentieth-century development of the former lead area. Smailes, in the early 1930's, showed that the percentage of agricultural holdings under fifty acres in size in the lead dales was far higher than that in the Cheviot dales. He also found that 'It is a really remarkable feature of social life in these dales how almost all of the labouring class families of old standing are found to keep a few cows in addition to their "regular" occupation. Lead mining has left as its heritage the tradition of part-time farming.'[60] In 1962 Thompson, studying the Durham dales, found that though there were still 'many holdings of uneconomic size', the number had decreased sharply since the 'thirties owing to amalgamations.[61] This trend seems likely to continue until most traces of the earlier mode of settlement disappear.

Gardens

By 1860 most miners without smallholdings had gardens. In Baker and Tate's *New flora of Northumberland and Durham* published in 1868, there are lists of the plants found growing at unusually high altitudes. The authors comment on the remarkably high ground on which they found cultivated plants in the lead dales, contrasting strongly with the meagre heights reached in the Cheviot dales. A smallholder at just below 2,000 ft in upper Teesdale was found to be growing rhubarb, potatoes and turnips 'in the hollow of a disused limekiln'. In Allendale, at over 1,600 ft, were found, in addition, cabbage, lettuce, onion, carrots and other vegetables, together with such fruits as plum, raspberry and currant bushes. So vegetables could be grown at these altitudes and in fact were. It is not clear, though, how long gardens had been commonplace, nor the proportion of miners who had an opportunity to cultivate one.

The earliest evidence on this point is contained in the returns to a questionnaire of the Poor Law Commission of 1834. One of the queries, directed to the officials of selected parishes, asked to what extent the cottages possessed gardens. In the Forest district of Middleton parish (upper Teesdale) the answer was 'None; gardens seldom pay for the seed and labour in this Siberian clime.' In Stanhope parish, about half the cottages were thought to have gardens, and in Alston parish 'very few'.[62] In 1842 Dr Mitchell said of Weardale that 'very often there is no cabbage-garden attached to the miner's cottage, but sometimes there is'.[63] This evidence, although scanty and unsatisfactory, does suggest that gardens were either non-existent or uncommon amongst the miners in the early nineteenth century, and if this were the case then it must also have been true of the eighteenth.

By the 1860's gardens seem to have been much more common. Bainbridge told the 1864 commission that 'in order to compensate such of our men as have not farms we endeavour to provide them with gardens'.[64] All the London Lead Company's cottages in Nenthead and Middleton had gardens attached.[65] Dr Peacock, in his evidence to the 1864 commission, remarked of the miners in Teesdale that 'there are comparatively few of them who have not the opportunity of cultivating a garden'.[66] The 1861 survey of the Beaumont estates in Allendale shows that, in the Coalcleugh area, of the nineteen houses without smallholdings, only five had gardens. In the Allenheads area the figure was much better; 34 of the 38 houses without smallholdings had gardens attached.

From the scanty evidence available, therefore, it appears that there was a decided increase in the number of houses with gardens during the nineteenth century, coinciding with the relative decrease in the number of houses with smallholdings or grazing rights on the commons. The enclosure Acts made it easier for land to be made available for gardens than in the eighteenth century.

Fuel

To obtain comfortable living conditions in the damp, cold climate of the lead dales, the inhabitants required a plentiful supply of fuel for their fires. Wood was too rare and valuable a commodity for household burning, so the main fuels were coal, peat and a poor-quality coal known as craw (or crow) coal.

Peat and craw coal were the usual fuels in the upper Pennine dales in the eighteenth century.[67] Coal from outside the district was used only when the weather made it impossible to obtain burnable peat. The Allendale smallholders whom Mulcaster mentioned in his letter about the enclosure award used their galloways to 'lead peats . . . which being their sole Firing except in a year such as the last before this was when they could not get them Dry were therefore with their Galloways obliged to fetch Coals from your Colliery at Stubbick, which is the nearest Colliery they have'.[68] In the nineteenth century roads improved, and the cart rather than the packhorse became the normal means of transport. By 1842 the inhabitants of the Durham dales— those nearest the coal mines—were abandoning peat for good coal.

On Saturdays in Weardale were to be seen 'many carts, conducted by miners, loaded with coal, for which they had gone to the nearest pit on the edge of the coal country'.[69] By 1872, according to a Weardale writer, coal was uniformly used in the dale, supplemented occasionally by a few peats.

Peat was dug all over the moors, and the enclosure Acts safeguarded the rights of the inhabitants to dig peats on the remaining unenclosed land. Their rights to obtain craw coal were also explicitly upheld in the Allendale and Weardale enclosure Acts. A nineteenth-century Weardale historian wrote an interesting account of how this poor-quality coal was mined and used 'thirty or forty years ago'. The miners cut tiny levels into the hillside to get the coal, which was manufactured into what were known as 'cats'. It was first crushed small and put on the floor of the 'cathouse', 'an outbuilding erected purposely to keep the cats in during the winter, frost being injurious to cat material'. With the coal a quantity of blue clay was mixed, in the proportion two parts of coal to one of clay. The two were then thoroughly mixed by a female 'cat treader', who trampled it for hours in heavy clogs. When a fire was lit it was started with dried heather and peats, then kept going with the cats, moulded into balls 'about the size of the common kitchen garden turnip'. Once the fire had been lit it was normally kept alight all through the winter.[70]

Lodgings

The journey to work was often quite a trek. The miners' homes were spread over the inhabitable parts of the dales; the mines were mainly at the heads of the rivers on the very edge of, or far beyond, the cultivable areas. The older villages—Alston, Allendale Town, Stanhope, Middleton—were a considerable distance from the main workings. The newer villages—Allenheads, Coalcleugh, Nenthead, Hunstanworth—were at the level mouths, but only Nenthead was of any size. Most miners lived in scattered dwellings, valuing smallholdings more than easy access to places of work. One Alston Moor miner questioned in 1842 described how 'I work at the mine Harehope Gill, three miles and a half off; I have to walk there and back in addition to my work. We have to descend a shaft by ladders from one stage to another; I have to go down near fifty fathoms by about twelve ladders altogether.'[71] This was a fairly typical distance for men who lived at home, though some walked much farther. Many others were forced to lodge away from home during the week, and return only at weekends.

In the eighteenth century most miners lived in accommodation near the mines during the working week. Bishop Pococke, in 1760, described the miners of Alston setting off on Mondays carrying 'their provision for the week to the mines'.[72] The habitable area at this time did not reach as far up the dales as in the nineteenth century. Mines tended to be small, and 'trials' at new places frequent. A 'shop' for the accommodation of miners was one of the first things to be built on the new site by large concerns and tiny partnerships alike.[73]

The development of the Derwent mines at Hunstanworth between

1842 and 1864 illustrates the growth of a mining settlement. In 1842 the chief agent of the Derwent mines told the commissioner that some of the workers lived in Blanchland parish, but that many came from 'Stanhope and Allendale, 10 miles off' and lived in lodging shops during the week. The commissioner himself remarked that around the mines 'the fell is altogether uninhabited'.[74] By 1864, however, most men lived near the mines, in the new village of Hunstanworth, and there were enough cottages in the district for lodgings to be provided privately for any who came from outside.

A settlement did not always grow up at an established mine. At the small Stonecroft and Greyside mine west of Hexham, near the Roman wall, in 1864, miners were 'what they call *wallet men*, that is men who carry wallets, from Alston and Allendale'. There was one 'shop' and a number of lodgings were available in the neighbourhood. The company had built two cottages near the mine for any 'wallet men' who cared to take them, but 'we have had a great deal of difficulty in letting them; the men prefer going elsewhere'.[75] In this matter of lodgings there were contrasts between the different dales, and at different periods within the same dale. In the two Allendales in 1842 and 1864 there were no 'shops' at all. Most men lived at home throughout the week, as the mine entrances were within the settlement area. For those who did not, plenty of lodgings were available in private houses. At Coalcleugh in 1864 'They have to go a short distance; they come down the country about two miles or so, and they get to their own homes at night.'[76]

Ten years later the mining area had changed, and men who had been conveniently placed for working in one mine had to lodge away from home to work at another, the old one having become exhausted. At Coalcleugh in 1871 'the mines which used to be so prosperous here, are now very poor, so that many miners have had to remove with their families, and others go and work all the week in other mines, and only come home from Saturday till Monday'.[77] This exhaustion—sometimes temporary, sometimes not—of an individual mine was a common occurrence. It discouraged many miners from moving house simply to be near their place of work, since there was never any guarantee of continued employment.

In Weardale and on Alston Moor, some mines were fairly close to inhabited areas, others were further away. Lodging shops were not necessary along the main length of Weardale above Stanhope, save at the very top of the dale and at isolated tributaries such as Rookhope

and Killhope. On Alston Moor the large settlement at Nenthead
supplied the labour needs of the biggest mine. Other mines were more
isolated. At Rodderup Fell, on the slopes of Crossfell, in 1864, 'many
of our men are living at considerable distances; some live at Alston,
and some up at Garrigill . . . but we have a rule that if they live more
than a certain distance from the mine they must lodge either at the
mine or in the neighbourhood'.[78]

The shallow valley of the Tees above High Force made settlement
impracticable around the most important mines in Teesdale. In 1864
more than two-thirds of the London Lead Company's labour force
there lived in lodging shops during the working week. Houses around
the mines were limited to the mining offices; the land was too barren
and exposed to permit any private settlement.

Three types of accommodation were available for miners living
away from home: (1) lodgings in private houses, (2) inns or boarding
houses, and (3) 'shops' specially constructed by the mining companies
for this purpose. Private accommodation was available only around
mines where there were settlements. Allenheads and Nenthead were
the prime examples. In 1842 'the usual price is 6d a week each, for
which there is a bed between two of them, leave to make their crowdy
on the fire in the morning, and they have their potatoes boiled for
them in the evening'.[79] By 1864 the normal price for a working week's
lodging had risen to 9d or 1s, the lodgers still being responsible for
finding their own food. The 1864 commissioners found that this system
was the main cause of any overcrowding that existed, but without
building large quantities of new houses the companies could do little
about it. The Rodderup Fell Company, however, had deliberately
constructed houses 'built larger than they would have been; they are
better houses, in order to facilitate the taking in of lodgers'.[80]

Private provision of accommodation for miners on a large scale in the
shape of proper inns or boarding houses was unusual. Even in the
eighteenth century, the owners were most unwilling for such houses
to be associated with the sale of drink. In 1766 one of the Blackett
agents requested 'that in case Sir Wr. Blackett will give him a piece of
ground anywhere in Killhope he will build a house upon it for his
nephew . . . to serve both as an alehouse & a house for miners etc., to
lodge in'. Sir Walter was inclined to grant the request, but insisted on
'his power to put a stop to selling Ale whenever he thinks proper under
forfeiture of the Lease.'[81] In 1842 and 1864 no arrangements of this
sort were in operation, at least in the Beaumont and London Lead

Company areas. All lodging houses were run directly by the mining companies—and in the case of the Lead Company the penalty for introducing alcoholic liquor was instant dismissal.

The mining companies established lodging shops in all the areas where there was an insufficiency of private accommodation. For the smaller mines the shop might be a building containing an office and a blacksmith's forge within two rooms, as well as sleeping accommodation. Many of the eighteenth-century shops must have been of this nature.[82] In the nineteenth century the two largest concerns built barrack-like buildings of two stories specifically to serve as sleeping accommodation for their miners. In both the 1842 and the 1864 reports there are many descriptions of these shops, the condition of which raised the ire of Assistant Commissioner Mitchell more than any other feature of lead mining life. (Plate 1 shows the outward appearance of a shop.)

He gives a precise description of one he visited in upper Weardale, presumably owned by the Beaumonts. It consisted of two rooms of identical dimensions, 18 ft × 15 ft, one above the other. The lower room had two windows in one wall, and a large fire at one end. Down the middle of the room was a long table and some benches. Along one wall were 48 small cupboards, some lacking doors, but most with them, and padlocked. There were also a number of cooking utensils. The upper floor, reached by a ladder, had no windows nor any other form of ventilation. Altogether there were fourteen beds, arranged bunk fashion, one above another. Each bed was about 6 ft long by $4\frac{1}{2}$ ft wide, intended for at least two persons, with possibly a boy sleeping across the foot as well. Other shops he visited were similarly designed.

Though the beds had not been occupied for three preceeding nights the smell was to me utterly intolerable. What the place must be in the summer's nights is, happily for those who have never felt it, utterly inconceivable. . . . I should think it no hardship to have to remain 24 hours in a mine, but I should be terrified at being ordered to be shut up a quarter of an hour in the bed-room of a lodging shop.[83]

The lodging shops do seem to have been especially bad; many of the statements of individual miners express as much abhorrence as the commissioner. 'The lodging shops are never washed; the beds and bed clothes are washed once or twice a year, being taken home by the men.'[84] Accommodation was free, the fire and the furniture being provided by the owners, as well as the building; the miners provided their own bedclothes and cutlery.[85]

By the 1860's the lodging shops of the larger companies had improved considerably, possibly as a result of the criticism in the 1842 report. Ventilation was adequate in all the bedrooms. By 1861 the London Lead Company whose shops seem to have been as bad as those of any other concern in 1842, had added 'A library and reading-room at each mine-shop, and it is one of the rules that a portion of Scripture be read, and prayers offered at the most convenient hour every evening.'[86] More materially, a flow of water was also available at all the company's shops, but two men to a bed was still the rule.

Their five-day week meant that the miners were able to have two complete days at home every weekend. They set off, with a week's provisions, early on Monday morning and returned on Friday evening. Some mines, where night shifts were worked, allowed their men to leave in the middle of Thursday night, and so have a long weekend. The washer boys worked on Saturday morning shifts, and so they had less time at home than their fathers.[87]

The settlement pattern in the lead mining area was based on a system of smallholdings. The altitude and the poor soil meant that there was no competition for land from full-time farmers. In other parts of England where a mining industry was located in poor agricultural land, as in the Yorkshire and Derbyshire Pennines, a similar system developed. In Cornwall there was an insufficiency of suitable land and by the nineteenth century, although some miners possessed smallholdings, most of them lived in villages close to the mine. The lodging shop was unknown there; if a miner was forced to live away from home, lodgings were always available in private houses near his work. The Durham colliers, working at mines situated in the middle of good agricultural land, lived in villages near the pit. In the nineteenth century the pressure of an increasing population on the limited area suitable for settlement forced more lead miners to live in villages, but they remained a small proportion of the whole.

NOTES

[1] A. E. Smailes, *The dales of north-east England*, 1932, p. 113.
[2] 1842 report (Mitchell), p. 722.
[3] White, p. 456.

[4] Raistrick, 1938, pp. 21–30.

[5] D. Thompson, *Rural geography of the west Durham Pennines*, 1962, pp. 68, 134–6.

[6] Sopwith, *Diary*, 29 March 1852.

[7] R. Pococke, *Northern journeys*. Surtees Society publications, vol. 204, 1914, p. 209.

[8] 1842 report (Mitchell), p. 722.

[9] R. W. Brunskill, 'Development of the small house'. *Transactions of the Cumberland and Westmorland Archaeological and Antiquarian Society*, new series, 53, 1952, pp. 160–89.

[10] Descriptions of habitations are given in: J. G. Nevin, 'The general system of farming in Alston', *Proceedings of the Durham College of Science Agricultural Students' Association*, vol. 1, part 1, 1900–01, p. 23; *Depressed condition of the agricultural interests*. Royal Commission report ('Durham' by Coleman), 1881, p. 228; Thompson, p. 132.

[11] Sopwith, 1833, p. 145.

[12] Adm. 66–96, 25 January 1769.

[13] 1864 report (appendix), p. 17.

[14] L.L.C. 31, 6 November 1866.

[15] Adm. 66–96, 24 February 1775.

[16] 1842 report (Mitchell), p. 736.

[17] *The state of the dwellings of the labouring classes in Cumberland, Durham, Northumberland and Westmorland*, pp. 409–44.

[18] Seventh report of the Medical Officer of the Privy Council, 1864, pp. 191–2.

[19] 1864 report (vol. 1), p. 261.

[20] Seventh report of the Medical Officer of the Privy Council, 1864, p. 180.

[21] 1842 report (Mitchell), p. 751.

[22] Poor Law Commission report, 1834, section B1–2–160b.

[23] *State of popular education in England*. Royal Commission report, 1861 (vol. 2), p. 366.

[24] Granger, p. 21.

[25] Eden, vol. 2, p. 169.

[26] Poor Law Commission report, 1834, section B1–3–160c, B1–2–98b; 1842 report (Mitchell), p. 757.

[27] Featherston, p. 15.

[28] *Depressed condition of the agricultural interests*. Royal Commission report ('Durham' by Coleman), 1881, p. 228.

[29] Poor Law Commission report, 1834, appendix A, p. 138A.

[30] 1864 report (vol. 2), p. 380.

[31] *Rating of mines*: select committee report, 1857, p. 16.

[32] Ibid, p. 26.

[33] D. Macrae, 'The improvement of waste lands.' *Journal of the Royal Agricultural Society*, second series, vol. 4, 1868, p. 323.

[34] Sopwith, *Diary*, 23 June 1866.

[35] Adm. 66–96, 20 April 1768.

[36] Adm. 66–96, 25 January 1769.

[37] Adm. 66–96, 6 September 1772.

[38] Adm. 66–96, 29 May 1772.

[39] Adm. 66–96, 31 August 1772; 13 June 1772; 1 March 1773.

[40] Raistrick, 1938, pp. 23–32.

[41] Thomas Bell collection.

[42] 1842 report (Mitchell), p. 757.

[43] Greenwich Hospital report, 1821.

[44] 1864 report (appendix), p. 327.

[45] Bailey and Culley, p. 200.

[46] Poor Law Commission report, 1834, section B1–1–98A.

[47] Hexham manorial papers.

[48] Dickinson, p. 258.

[49] 1842 report (Mitchell), p. 748.

[50] Thomas Bell MSS.

[51] *Teesdale Mercury, Tales and traditions, c.* 1885, p. 164.

[52] The chief parishes involved were enclosed as follows:

(a) *Hexamshire and Allendale commons:* date of Act, 1792; date of award, 1800.

(b) *Weardale* (mainly Stanhope parish): date of Act, 1799; date of award, 1815.

(c) *Alston and Garrigill:* date of Act, 1803; date of award, 1820.

(d) *Middleton:* date of Acts, 1804 and 1834; date of awards, 1816 and 1841.

[53] L.L.C. 16a, 13 April 1811.

[54] Thomas Bell MSS.

[55] Sopwith, *Diary*, 16 August 1856.

[56] Ibid, 6 August 1856.

[57] Macrae, pp. 321–34.

[58] *Depressed condition of the agricultural interests.* Royal Commission report ('Durham' by Coleman), 1881, p. 228.

[59] Dickinson, p. 2.

[60] Smailes, appendix IV and p. 125.

[61] Thompson, Fig. 32.

[62] Poor Law Commission report, 1834, sections B1–2–98B/149B/160B.

[63] 1842 report (Mitchell), p. 748.

[64] 1864 report (vol. 2), p. 380.

[65] Raistrick, 1938, pp. 31–2.

[66] 1864 report (appendix), p. 18.

M

[67] Bailey and Culley, p. 168.

[68] Thomas Bell MSS.

[69] 1842 report (Mitchell), p. 722.

[70] W. M. Egglestone, *The Weardale nick-stick*, 1872, pp. 85–8.

[71] 1842 report (Mitchell), p. 761.

[72] Pococke, p. 209.

[73] B/B 133, 22 August 1753.

[74] 1842 report (Mitchell), pp. 722, 758.

[75] 1864 report (vol. 2), p. 355.

[76] Ibid, p. 539.

[77] *Spar from the high flat*, 1871, p. 31.

[78] 1864 report (vol. 2), p. 339.

[79] 1842 report (Mitchell), p. 740.

[80] 1864 report (vol. 2), p. 339.

[81] B/B 48, 25 June 1766.

[82] J. W. Allan describes a shop like this at one of the smaller mines on the slopes of Crossfell as late as the 1870's: *North-country sketches*, p. 46.

[83] 1842 report (Mitchell), pp. 740–2.

[84] Ibid, p. 770.

[85] 1864 report (appendix), p. 20.

[86] *State of popular education in England*. Royal Commission report, 1861 (vol. 2), p. 373.

[87] 1842 report (Mitchell), pp. 740, 758; 1864 report (vol. 2), p. 355.

VIII

The supply of food

In spite of the part-time agricultural activities of the lead miners the mining region was not self-sufficient with regard to food. Its altitude prevented the growing of cereals, and despite its rural appearance it was an industrial area into which most foodstuffs had to be brought from outside.

In 1797 Eden published three examples of the expenses of lead miners, one of a family in Weardale, the other two of families in Nenthead.[1] These show an interesting uniformity in the relatively high amount of expenditure devoted to food. The Weardale family had a gross expenditure the previous year of £34 14s, of which £27 2s was spent on food. The first Nenthead family spent £34 10s out of £44 on food, and the second £40 out of £48. In a similar budget of the family of a man employed by the London Lead Company, the 1842 report cites a weekly expenditure of 7s 7d on food out of a total of 9s 1¾d.[2] These figures suggest that about 80 per cent of each family's income was spent on food. Eden also published what he described as 'the usual annual expenditure . . . of an agricultural labourer in the county of Cumberland'.[3] This quotes a total annual expenditure of £18 18s 6d, of which £13 4s 6d, or slightly less than 70 per cent, was devoted to food. The labourer quoted appeared to buy all his food, and not to receive any in lieu of wages, as was usual in the parts of Northumberland where the 'bondage system' operated.

It would be a mistake to draw too many conclusions from these notoriously unreliable 'specimen' budgets, but the figures do suggest that lead miners were forced to spend a higher proportion of their income on food than the neighbouring agricultural labourers. The altitude of the lead mining dales largely prohibited the large-scale growing of any vegetable food. Much of it, and all cereals, had to be imported from the arable lands further down the dales, and until the coming of the railways, transport costs kept prices high. At the end of the eighteenth century food prices in the rural north of Northumberland, where more was produced than was needed for local consumption, were among the lowest in England; in the south-west of the county grain prices at Hexham market were higher than in the city of Newcastle![4]

The cost of food is to a large extent dependent on the type of food-stuff bought, meat for example being relatively expensive, cereal foods relatively cheap. Unfortunately, it is not easy, from the nature of the evidence available, to give a precise history of the content of a lead miner's diet in the eighteenth and nineteenth centuries, or of the changes and developments in diet during that period. There is vir-tually no evidence before the middle of the eighteenth century, except some rather unreliable memories recounted in such works as Eden's *State of the poor*. For the remainder of the period sporadic information is available.

One important factor that the Government commissioners empha-sised strongly in both 1842 and 1864 was the deleterious effect of the poor ventilation in the mines on the appetites of the miners. An owner of one of the smaller mines on Alston Moor stated in his evidence to the 1864 commission, 'I have observed . . . that when a shepherd comes in off the fells to take refuge from the bad weather and to get his supper in a miners' hut, he will eat at a meal as much as four or five of the miners will eat; their appetite, generally speaking, being so much smaller.'[5]

A working miner who migrated to the coal areas in 1831 but subsequently returned, remarked in 1842 that one of the advantages of being a lead miner was that 'We can work . . . with less food, and food of an inferior sort from what is required in the coal mines; it is canny work rather than hard; our mines do not excite an appetite like the coal mines.'[6] The same miner also noted that the washer boys had much better appetites than their fathers who worked inside the mines.

Bread

Bread and other cereal foods formed much the greater part of a lead miner's diet throughout the eighteenth and nineteenth centuries. The grain used to make bread appears always to have been the cheapest available at the time. Writing to Sir Walter Blackett in 1772 about the bread eaten in and around Newcastle, Henry Richmond, his chief agent, divided bread into four classes according to its expense. The first and dearest was 'fine wheaten Bread', which was eaten by the upper classes, and by 'the Keelmen and some of the other Workmen in the Coal Works whose Great Earnings set them above Eating any but the finest Wheaten Bread that is made'. The second class was wheaten bread of an inferior quality, the third maslin (a mixture of wheat and rye), and the last, rye bread.[7] In a subsequent letter Richmond said that the bread of the miners was 'mostly Rye'.[8] This appears to be the earliest direct reference to the kind of bread eaten in the lead districts, but in the Blackett/Beaumont accounts there are records of the purchase of foreign rye for use by the miners as early as 1728.

From at least 1724 until 1740 corn was regularly bought in Newcastle for consumption by the lead miners in the employment of the Blackett family. In the ledgers recording its purchase it is usually described simply as 'corn', with no mention of its nature or quantity. But for two successive entries in the accounts for 1734–35, the figure is broken down. In the half-year July to December 1734, the miners of Allenheads and Coalcleugh were supplied with 424 bushels of wheat, 2,954 of rye, 340½ of pease and 146 of beans. Thus rye formed about 75 per cent of the food despatched in this period. In the nine months January–September 1735 Allenheads, Coalcleugh and Weardale were sent 531¾ bushels of wheat, 7,112½ of rye, 1,235¼ of pease, 426½ of beans and 796 of barley. Rye then formed about 70 per cent of the total. Rye would appear, therefore, to have been the most common bread cereal, at least in the Blackett mining areas, in the early part of the eighteenth century.

At the beginning of the nineteenth century, rye bread was described as 'the most general bread of the labouring people' in the southern part of Northumberland.

After being leavened, until it gains a considerable degree of acidity, it is made into loaves, and baked in a large brick oven, or made into thick cakes, called 'sour-cakes' and baked on the girdle: the bread is very firm and solid, dark coloured, and retains its moisture or juiciness longer than any other bread we know.[9]

Rye does not seem to have been very extensively grown in Cumberland, where the usual bread at this time was made of barley, but Alston received most of its food supplies from Northumberland. Even in Northumberland the cultivation of rye was declining; Bailey and Culley record that although it was formerly the principal grain crop of the county, by the beginning of the nineteenth century it was grown only on soil suitable for nothing else.

In 1794 the Beaumont chief agent wrote to the Coalcleugh agent that one of the contributory causes of the scarcities of that year was that 'the great Importation of Rye at this Port [Newcastle] in 1790 and 1791 reduced the Price so low as to discourage the growth of that article, not only in this country but in the Southern Counties, where instead of growing Rye as formerly they now grow Turnips'.[10] While this statement may not have been entirely true, it is extremely interesting in its suggestion of the increasing difficulty of obtaining *English grown* rye by the end of the eighteenth century.

In 1825 the bread of the Allendale lead miners was still made 'mostly of Rye',[11] and some of the witnesses to the 1842 commission still ate rye bread, but a comment by one miner suggests that it was no longer the usual diet: 'The London Company, in 1839 and 1840, imported a great deal of rye-corn, and had it ground and sold to the workpeople. The bread did not agree with many, and those who could possibly do without it gave it to the pigs.'[12]

The county surveys printed in the *Journal of the Royal Agricultural Society* in the 1840's and 1850's scarcely mention rye as a crop in the northern counties, and no example of rye on its own as a bread grain is mentioned by Edward Smith in his report to the medical officer of the Privy Council on the food of the poorer labouring classes in 1863. Hence although it may well have been familiar in the lead dales when eaten nowhere else in England, rye bread appears to have been forsaken there shortly after 1841.

Maslin bread was being used rather later than bread made of rye alone. Bailey, writing about 1800, stated that it was the usual bread of County Durham,[13] and in 1834 it was the main food in Weardale and Teesdale.[14] By 1862 Smith found that it had almost vanished from the labourers' diets, but he quotes some examples of farm labourers at Haydon Bridge who ate bread consisting of an equal mixture of maslin and wheaten flours.

The grain that appears to have replaced rye as the main constituent in the miners' bread was barley—at a time when its use as a breadstuff was dying out elsewhere in England. Barley had been the usual bread grain in Cumberland and north Northumberland throughout the eighteenth century, and in 1797 Eden listed it in his budgets of lead miners in Weardale and Nenthead. In the bread shortages of the 1790's the Beaumonts purchased barley as well as rye for distribution to their workmen. It is mentioned in the evidence to the commission of 1842. In 1862 Smith found that the use of barley bread had much diminished 'even in Northumberland'[15] but it was evidently still the main bread-stuff of the miners as late as the 1870's.[16]

Wheaten bread, almost universal in the south of England by 1800, and in the north by 1850, is scarcely mentioned in any of the sources relating to the lead miners' food. Its price was always too high, even when the construction of the railways brought down transport costs. Allan records a conversation with Allendale lead miners, some time after 1870, in which they passionately affirm the superiority of barley over wheaten bread. Conservatism in eating habits is always strong—particularly when reinforced by financial stringency.

Oatmeal was apparently never used as an actual breadstuff, as it was in parts of Westmorland in the eighteenth century, but in the form of 'crowdy' or 'hasty pudding' it was more frequently eaten than bread itself, right up to the closure of the mines in the 1880's. Both were forms of porridge, eaten with milk, treacle or butter. Purchases of oatmeal formed a substantial part of the expenditure of all the miners whose budgets are given by Eden, second only to purchases of rye and barley. According to the 1842 report all the miners ate great quantities of crowdy. A miner stated that 'one man who worked along with me lived for three years on oatmeal and water, and never had anything else, unless the other men gave him bread'.[17]

The reasons for this great consumption of oatmeal were that it was both cheaper than bread and more convenient to prepare for men who frequently lived away from their homes for four or five days a week. The report of the 1864 commission shows that oatmeal was then still as fundamental a part of the miners' diet as it had been in 1842, although its use had declined greatly in such areas as Derbyshire, where it had previously been an equally important item in the labourers' diet.[18] In 1878 the Commission on the Agricultural Distress was told that oatmeal was still the staple food of the agricultural labourer in the extreme north of England—nearly a century and a half after Dr Johnson's famous definition of oats, which limited their *human* consumption to Scotland.

In the eighteenth century the bread of lead miners was made chiefly of rye and maslin. Rye was used in Allendale and Alston Moor, which were supplied from Northumberland, and maslin in Weardale and Teesdale, which were supplied from County Durham. In the early nineteenth century these were gradually superseded by barley bread, when less and less rye was grown in England, although in times of scarcity rye was still imported from the Baltic by the employers.

Oatmeal was eaten throughout the eighteenth and nineteenth centuries, and there is no evidence that wheaten bread ever formed a staple part of the miners' diet.

The factor deciding the type of grain eaten was its price. In the eighteenth century (so far as comparative statistics are available), wheat was about twice as costly at market as barley, rye cost slightly less than barley, and oats were cheaper still. For example, in 1767 London prices per quarter were: wheat, 40s–51s; barley, 23s–27s; rye, 22s–23s; oats, 13s–15s. By the end of the eighteenth century, rye and maslin had become more expensive than barley at northern markets,[19] and this was probably the decisive factor in enforcing a change from rye to barley in the lead districts. Throughout the nineteenth century wheat was far more costly than barley, which in its turn was more expensive than oats.[20]

In the eighteenth century the miners' bread was little different from that of the bulk of the agrarian population of northern England but by the middle of the nineteenth it was decidedly old-fashioned. Sir William Ashley, in his Ford lectures for 1923, noted that although a writer in the south of England discussing agriculture in 1764 thought maslin bread was extinct, its price was 'regularly quoted in the returns of the Hexham market as late as 1841'.[21] In his article on bread in the second edition of his *Dictionary of commerce* (1840), J. R. McCulloch referred to Eden's statement that wheaten bread was virtually unknown in Cumberland, and added:

Everyone knows how inapplicable these statements are to the condition of the people of England at the present time. Wheaten bread is now universally made use of in towns and villages, and almost everywhere in the country. Barley is no longer used, except in the distilleries and in brewing; oats are employed only in the feeding of horses; and the consumption of rye bread is comparatively inconsiderable.

The lead miners' bread may have had as much, or more, nutritious value as the wheaten loaf but it was certainly primitive by Victorian standards.

Other food, and drink

Potatoes were a very important item in the lead miners' diet. They grew in the gardens of cottages at the highest altitudes and, where those grown at home were not sufficient, were supplemented by purchase.

All the mining families whose budgets were printed by Eden bought potatoes. This was at the end of the eighteenth century, when they had been an established article of food in the north of England for at least half a century. In 1800 there was a Durham village where potatoes had been grown as 'an article of trade' and their cultivation had been 'the principal employment of several families for upwards of eighty years'.[22] In the nineteenth century it is clear that potatoes were almost as important a food as bread itself. During the recession of the early 1830's a return to the Poor Law Commission of 1834 described the Alston miners as 'subsisting chiefly on potatoes, with a little oat-meal'.[23] Nearly every lead miner questioned in 1842 named potatoes as forming the chief meal of the day, sometimes with meat but normally without it. During the shortages of 1794 the Beaumont chief agent told the Coalcleugh agent to 'recommend to [the miners] . . . that they make use of a mixture of Potatoes with the Rye flour agreeable to the enclosed Receipt, which makes wholesome good bread.'[10] Whether the miners heeded this recommendation is not recorded, but in normal times they preferred their potatoes either boiled or made into a form of pie, not mixed with flour.

Other fresh vegetables were probably consumed according to whether the miners possessed gardens or not, and only occasionally purchased. In the 1790's the Allendale market was supplied with fresh vegetables from the Hexham area, but the amount was inconsiderable compared with what was sent to Newcastle.[24] This contrasts with Hexham's grain crops, most of which at this time went to Allendale and Alston Moor. Fruit and vegetables could be grown at remarkably high altitudes in the lead dales but full advantage was not taken of this fact until well into the nineteenth century. In the eighteenth and early nineteenth centuries gardens were rare, and green vegetables presumably equally uncommon, for there would be little money to spare for their purchase, and they were not regarded as an essential article of diet. By the mid-nineteenth century, as we have seen, the mine owners were actively encouraging gardening, and most families grew their own vegetables or had access to a local supply some time during the year.

Milk, butter and cheese were similarly available to families which possessed cows or goats. Nearly all the men and boys questioned in 1842 drank milk every day with almost every meal. Dr Peacock, one of the medical witnesses who appeared before the 1864 commission, thought that the northern lead miners were 'certainly a more robust race' than the Cornish ones, and that this was largely due to the greater amount of animal food in the form of milk that the northerners consumed.[25]

Fresh meat was a luxury that, except in prosperous times, could be afforded only at irregular intervals. Bishop Pococke, who visited Alston at a time when the lead trade was enjoying comparative prosperity in 1760, observed:

... they have great markets here for meat every Saturday. From Christmas to Easter they kill weekly twenty calves and four beeves; from Easter to Midsummer 50 calves and 6 or 7 beeves; from that to the first of september 20 sheep and 40 lambs; for six weeks before Christmas 30 beeves and 20 sheep, being the time they lay in salt stores of beef; and at Christmas, 'tis said, they have been known to kill 17 beeves, 500 sheep, seventy calves, and a 1000 geese.[26]

In less affluent times, meat consumption was very much lower. In 1834 the return to the Poor Law Commission relating to Upper Teesdale stated succinctly, 'They get not Beef now.' Even during better times, most miners would have fresh meat only once a week, and on holidays. Many of the men questioned in 1842 and 1864, if they ate fresh meat at all, did so only on Sundays. In 1842 bacon was the meat most frequently mentioned and 'butcher's meat' was but an occasional delicacy. Goose was the favourite luxury dish in Allendale until at least the end of the nineteenth century.[27]

Other items of food occasionally varied the diet. Salmon and trout from the rivers, grouse, rabbits and hares from the moors were there for the poaching. A number of miners kept bees. Bilberries and blackberries grew wild.

'Their chief beverages are water and tea' was said of the drinking habits of Allendale miners in 1825.[28] This remained generally true for the rest of the nineteenth century in all the mining areas. The Government Commissioners in 1842 and 1864 commented on the comparative absence of alcohol from the miners' diet. In the eighteenth century, lead miners were not renowned for their temperance, but the growth of Methodism and the increasingly puritanical regulations of the largest mining concerns effectively eliminated drunkenness from the region. Tea took the place of alcohol as a stimulant and was probably drunk as a substitute for milk by families who had no cow or goat. Tea was purchased by all the families whose budgets were printed by Eden at the end of the eighteenth century, and was mentioned by most of the miners questioned in 1842, although by this latter date many preferred coffee.

The necessity for a large proportion of the miners to live away from home for four or five nights a week affected diet. Food had to be such as would remain fresh and could be prepared by the most primitive cooking methods. A washer boy employed by the London Lead Company described his week's food in 1842:

I went on Monday morning, and took with me bread, white and brown, potatoes, bacon collop, half a pound, coffee an ounce, and sugar half a pound, and that supplied me till Saturday at twelve. I took with me on Monday three pints of milk, and we kept it up the level a little to preserve it sweet.[29]

The only cooking done at the lodging shops was the preparation of crowdy for breakfast, and of potatoes and sometimes bacon for supper. However, the men who lived at home seemed to eat in much the same manner. Typical was another London Lead Company washer boy: 'I get crowdy and milk to breakfast; I get 'tatoes to dinner, and salt, sometimes butter; I take tea for supper and eat bread with it.'[30] Another miner said, 'A great many miners have oatmeal crowdy to breakfast, which is cheaper than bread. Men who have no cow have no milk to the crowdy, they have only a bit of sugar or a bit of butter.'[31] Lunch, taken at about eleven or twelve o'clock, was normally eaten inside the mine and consisted of bread, and possibly a little bacon. The evening meal was again bread with hot potatoes. A better meal, with meat if it could be afforded, was eaten on Sundays.

The surviving evidence relating to the miners' food is mainly nineteenth-century, with a few scraps of information from the eighteenth. What is available, however, does suggest that changes in diet in the century between 1750 and 1850 were less fundamental in the mining area than elsewhere in England. At the beginning of the period the diet of the lead mining families was backward compared with the condition of labourers in the south of England, but not as compared with those in the agricultural north. At the end, however, the lead miner's diet was backward even compared with that of his agricultural neighbours, and when comparisons are made with the neighbouring pitmen of Northumberland and Durham it is seen that the lead miners had always fared worse for food. In 1772 the pitmen ate the most expensive bread, the lead miners the cheapest. One lead miner questioned in 1842, who had also worked in the coal mines, said, 'We do not eat so good victuals, nor near so much, nor do we drink so much beer, because we cannot afford to do these things.'[32] By this time the colliers ate only wheaten bread. In Cornwall, however, the miners' diet appears to have been as impoverished as that of the northern lead miners. Barley bread, scarcely any meat, and a heavy reliance on potatoes were reported in 1842,[33] and in 1864 one medical witness thought that the northern miners ate 'altogether more fresh animal food than the Cornish miners ordinary have.'[34] The Cornish men did have the decided advantage of plentiful supplies of fresh and salted fish, unknown in the northern Pennines. Moreover the Cornish mines were situated in food-producing areas. In the north, by far the greater quantity of food had to be imported into the mining region from some distance away.

Marketing

Although many of the miners possessed smallholdings, this source could not supply all their needs. Their animals would provide milk and occasionally meat. Their gardens—which were not common until well into the nineteenth century—would yield potatoes, and a certain amount of other vegetables and fruits. But basic foodstuffs—the cereals —had to be supplied from elsewhere. The full-time farmers grew corn at higher altitudes in the lead dales than they do today, but even so the entire mining population could not be supplied from nearby sources. The existence of this hungry population was very important to the farming economy of Northumberland and Durham. In 1794 the Durham lead mines made 'a considerable difference in all the farms at a great distance as the farmer is always certain of a ready market for every article his farm produces, and at the best prices.'[35] A farmer from Anick Grange (north of Hexham), questioned in 1829 as to the utility of the proposed Newcastle & Carlisle Railway, said that he would not send his corn to Newcastle as he could obtain better prices by sending it to the lead districts.[36] At an earlier date, 1797, it had been argued that the proposed canal between Newcastle and Haydon Bridge could be used to carry rye *from* Newcastle for the 'poor and labouring persons' in the lead mines.[37]

The various agricultural surveys and directories of the northern countries give a fairly clear picture of the marketing arrangements in the lead mining districts before the coming of the railways. Large market towns in the farming land at the foot of each of the lead mining dales acted as suppliers to smaller markets within the actual mining areas. Hexham supplied both Allendale and Alston Moor, Wolsingham supplied Weardale and Barnard Castle, Teesdale.

Alston, although in Cumberland, was more closely connected with Northumbrian agriculture because the roads westward over the Pennines were very poor until the 1820's. The Hartside pass on the road (or track) to Penrith was impassable to carts in the eighteenth century, and any food brought into Alston from that direction had to be carried on packhorses.[38] Very little, in fact, was so brought for the road to Hexham was appreciably better, and it was thence that most of the food came to supply Alston's Saturday markets. By 1829, while Hexham remained the main supplier, more food came from the areas immediately west of the Pennines and around Brampton, which had been made more accessible by new Greenwich Hospital roads. When railways were constructed, the Haltwhistle–Alston branch, completed in 1852, encouraged the inhabitants of Haltwhistle to expand their market to act as suppliers to the Alston Moor mines.[39]

Market day in Allendale Town was Friday. The town was sufficiently central to be reached from both Allenheads and Coalcleugh, and there were no secondary markets in those places. It was supplied exclusively from Hexham. After the construction of the Hexham–Allendale railway line in 1858, the market in Allendale rapidly declined, the inhabitants presumably preferring to make their purchases in Hexham. By 1886 it existed 'only in name'.[40]

Weardale was larger and its markets were more numerous. The Tuesday market in Wolsingham was the biggest, supplying the smaller centres, and although east of the lead mining district it was sufficiently far up the dale for many miners to attend in person, even in the eighteenth century.[41] There was a sub-market at Stanhope, in the middle of Weardale, on Fridays, and a further one near the head of the dale, in the heart of the actual lead mining district at St John's Chapel, on Saturdays.

The Teesdale mines were supplied from the Wednesday market at Barnard Castle. 'Corn is sold higher here, than in any market in the County, for the consumption of the mining districts to the westward' wrote Bailey in 1809.[42] There was a sub-market every Saturday at Middleton in Teesdale.

Corn was sold to the miners as grain or as flour. In 1829 it was the millers who bought the grain at Hexham market to send it to the lead mining districts as flour.[43] In the eighteenth and early nineteenth centuries, however, the miners also bought grain themselves, and had it ground at the corn mills which were to be found in all the mining dales. The millers were highly unpopular in times of shortage, as their customers blamed them for the high price of flour. In a corn riot at Wolsingham in 1795 lead miners destroyed the corn cylinders at the local mills.[44] All the corn imported into the area by the mine owners was delivered to the miners in the form of grain. By 1840, in Weardale at least, the local miller supplied the grain, which his customers bought as flour.

Credit

The industry's peculiar system of payment more or less forced the miners to live on credit. An unlucky year left them owing instead of receiving at the pays. In 1840 the Burtreeford (Weardale) millers were praised by a local writer for their generosity in supplying flour to the unfortunate on credit, to be repaid 'when fortune changed in their favour'.[45] The local retailers adapted themselves to the conditions in the mining industry. The weekly markets were larger and better attended immediately after the payment of subsistence. Great fairs were held during the pays in the eighteenth and early nineteenth centuries.

By the mid-nineteenth century both the larger mining companies were trying to discourage their employees from relying too heavily on credit. Thomas Sopwith told the Allenheads miners in 1846 to use their increased subsistence money to pay cash instead of living on credit, as he calculated they were charged 'thirty to thirty-five per cent more for credit'.[46] More practically, the London Lead Company had already established 'ready money shops' in Middleton and Nenthead, where the lessee was enabled to make use of the company's transport facilities to obtain supplies, on condition that he refused to give credit.[47] In all the lead mining centres the number of shops increased during the nineteenth century, as is shown by the lists in directories of different dates.

Subsidised corn

The poverty of the mining communities, together with the extra cost added to food prices by transport charges, meant that they were unable to support themselves all the time, particularly in times of national shortage due to bad harvests. Apart from humanitarian motives, the owners and lessees of the mines found it to their own advantage to give their employees help in such times. A wholesale exodus would have been forced on the miners were no help forthcoming, and the owners did not wish to lose their labour. Until the new railways and better roads had lowered transport costs and brought prices in the area down to a more normal level by the mid-nineteenth century, the owners assisted their employees in various ways. The records of the Blackett/Beaumonts, the London Lead Company and the Greenwich Hospital all have spasmodic references to help for their employees in obtaining cheaper food. Such aid took different forms at different times—regular wholesale importation of food into the district to supply quite a large proportion of the grain needed; occasional supplies in times of acute shortage; and encouraging the miners to form their own co-operatives, or corn associations, so that they could buy corn in bulk from Newcastle.

In the early eighteenth century the Blacketts sent large quantities of grain to their miners in Allendale and Weardale regularly. From the earliest extant accounts in 1724 until 1740, sums of money are recorded in the accounts every year as paid for corn purchased in Newcastle, or abroad, to be distributed to the miners at cost price. The figures cover unstated periods, but until 1736 between £500 and £1,000 appear to have been spent in this way annually. From 1736 till 1740 the payments diminish, and no record of any more can be found between 1740 and 1754. The amount of money involved suggests that the Blacketts were obtaining and distributing a sizable proportion of the grain their miners needed. Certainly, in the years of occasional scarcity later in the century, the sums involved rarely approached the annual £500–£1,000 of the earlier period.

During the century following 1740 the mining companies sent grain to the mines only in times of acute shortage, often at the miners' request or to avert threatened strikes. During the poor harvests of the 1750's, culminating in the disastrous one of 1756, corn was sent to the Blackett mines, and in 1757 a total of £1,041 4s 11d was spent on rye imported from Danzig.[48] No record can be found in the Blackett/Beaumont accounts of corn again being sent until 1783 (1782 was another very bad harvest) when small quantities were sent to Coalcleugh, and probably to the other mines as well. In the same year the Greenwich Hospital also supplied grain to its workmen commencing the Nent Force level.

In the mid- and late 1790's there was a series of bad harvests and consequent shortages. The correspondence of the Beaumont chief agent for these years mentions numerous petitions from the workmen for corn. In December 1794 he received a letter signed by more than a hundred Allendale miners complaining of shortages, and alleging they were being given short measure in Hexham market and by local millers. He wrote to the Coalcleugh agent that he was buying corn in the south of England that should arrive the following February; this should be distributed to the miners 'in proportion to the Number of their respective Families taking care that they have the legal Measure and are not imposed on by the Millers'.[10] This was done, but a year later the chief agent wrote to Colonel Beaumont:

Notwithstanding what you have done towards the relief of the Miners and Smelters during the time of the high price of Corn (which example neither the Lead Company nor any other Proprietor have followed, that I have heard of) the Miners and Smelters of your Works stopt the Works and some of them . . . committed some depredations by seizing a Cart with Flour and Oat Meal which they disposed of at their own price.[49]

The reference is to the riot at Wolsingham. The grain supplies continued, however, both before and after the Weardale miners' strike of 1797. The men were not satisfied with the quantity of food supplied, and in December 1799 the chief agent wrote to Colonel Beaumont:

Should you have any Application made to you by the Weardale Miners about Corn, they really should not be listened to, for they are the most dissatified, troublesome Workmen that you employ and have always been the foremost Promoters of any Mischief that is going forward. I send the Rye up to Weardale as fast as I can get the Carriers to take it, but not content with that they are perpetually coming down, some for two Bushels, others for four, which cannot be complied with.[50]

In 1800 there are records of action taken by the other two big concerns. In February the Greenwich Hospital receivers ordered the agent at Langley smelt mill to find out 'what is doing by the Governor & Co., Colonel Beaumont, etc., for relief of the necessitous at this Time when Corn & all necessaries are so extravagant & dear'.[51] In 1796 the Langley smelters had been given a temporary wage increase on account of high food prices.[52] This time, however, the smelters and miners employed in the Nent Force level were sent corn from Newcastle, supplies continuing until after the good harvest of 1801.[53] In June 1800 the London Lead Company spent £500 on corn for its miners in Teesdale and Alston Moor.[54] In the same year the company purchased an old lead mill near Garrigill and refitted it as a corn mill, the new lessee being instructed to charge fair prices and to produce good flour.[55]

After 1801 corn prices dropped, and the lead industry entered a temporary phase of prosperity. By 1808 the demand for lead was again falling, but it was not until 1816, after another disastrous harvest, that the men renewed demands for corn supplies. On 26 December 1816 the Beaumont chief agent wrote to Colonel Beaumont:

With regard to supplying the Miners with Corn from Newcastle, the subject had full consideration at the time, and from the little satisfaction which was expressed by them on a former occasion, it was not deemed expedient to interfere with the various establishments and dealers in Corn that supply them, and who in receiving their Monthly subsistence *in Money* could it was supposed purchase on equally good terms, whatever their consumption required.

Four days later, however, after a meeting with representatives of Weardale miners, he had changed his mind. Rye was to be purchased in Newcastle, as 'that which they were supplied was both very dear and unsound'. Similar complaints had been made in Allendale, so corn would also be sent there, 'cheaper and of better quality than can be obtained in the Country'.[56] The London Lead Company, too, sent rye and oatmeal from 1817 to 1820.[57]

During the 1820's and early 1830's, no grain appears to have been despatched by any of the mining concerns. The severe depression around 1830 forced many lead miners to leave the area; it is probable that no help was given this time, as superfluous labour was not wanted. In 1838, when trade had picked up, the London Lead Company sent a quantity of grain from Newcastle in response to a request from its miners.[58] This was apparently the last time that any company did so; better transport had reduced the cost of grain in remote areas relative to the cost in agricultural centres. The normal channels of trade were now considered adequate for providing the lead miners with corn.

Foreign corn was nearly always purchased when the corn laws permitted its importation. Rye, in particular, was far cheaper abroad. Details of the administration of the distribution of corn to the work-men are rarely mentioned in the records. The Beaumonts tried to distribute the corn in Allendale in 1795 'in proportion to the Number of their Respective Families'.[10] The Greenwich Hospital receivers laid it down in 1800 that 'We know that it is impossible to make a distribution that is in proportion to want or even to earnings, but the Workmen must take it as it is settled.'[59] The miners were generally charged the Newcastle bulk price for their grain, with nothing added

N

for transport costs, and in 1816 the London Lead Company charged its miners less than the cost price of the rye.[60] Money owing for the corn was either deducted from earnings at the pays, as appears to have been the Blackett/Beaumonts' practice in the eighteenth century, or corn was given in lieu of subsistence allowance.

Co-operation among the workmen themselves to obtain corn from Newcastle at less than the local rate is first mentioned in a letter from the Greenwich Hospital receivers to the agent at Langley mill.

We have no objection to John Friend getting Corn for the Nent Force Level Workmen from Newcastle, but as we told him when he was here in December we wd. have nothing more to do with the furnishing the Corn ourselves. . . . We however have no objection to their continuing to send to Newcastle for Corn as long as they find it any particular convenience to them (the workmen) to have it from there.[61]

It would appear quite probable that other miners employed in relatively small operations on Alston Moor might have done the same. At a later date the London Lead Company deliberately encouraged its workmen to organise themselves in corn associations. These were formed in the 1840's, and were allowed to make use of the company's transport facilities. Control of the corn mill near Garrigill was handed over to them.[60]

Starvation and malnutrition remained very near threats to the population of the lead dales until at least after 1850. A rise in national food prices or a decline in the demand for lead meant immediate hardship. 'The price of provisions are so high, that many Families in this poor Neighbourhood are half starved,' wrote a London Lead Company agent in 1810.[62] The population depended absolutely on work in the mines, and if circumstances were unfavourable it was a choice between starvation and emigration. In 1796 a woman whose husband had been dismissed from his work as a lead carrier for signing a petition wrote to Mrs Beaumont:

Deiar ladey . . . I humbly Beg that you will help us in our neads we having no oder way to get our Bread; and now we have no way; and do not know which way to persue for work to mentane our family. deir madem . . . in the name of god I hope you will take pity on our Small family that hath nothing Independent but what we must indever for; we farm A pese of ground; But it doeth not grow one Shef of corn to mentain them; and we having very little money to take at your pay we cannot pay our Creadet. I feir our little stok be taken from ous, if we get no mor worke under you.[63]

NOTES

[1] Eden, vol. 2, pp. 87–9, 170.
[2] 1842 report (Mitchell), p. 761.
[3] Eden, vol. 2, pp. 104–7.
[4] Bailey and Culley, pp. 166–7.
[5] 1864 report (vol. 1), p. 780.
[6] 1842 report (Mitchell), p. 762.
[7] B/B 48, 11 December 1772.
[8] B/B 48, 1 March 1773.
[9] Bailey and Culley, pp. 79–80.
[10] Blackett of Wylam papers.
[11] Mackenzie, 1825, vol. 1, p. 207.
[12] 1842 report (Mitchell), p. 762.
[13] Bailey, 1813, pp. 124–5.
[14] Poor Law Commission report, 1834, sections B1 and B2.
[15] Sixth report of the Medical Officer of the Privy Council, 1863, p. 240.
[16] Allan, pp. 60–5.
[17] 1842 report (Mitchell), p. 772.
[18] Sixth report of the Medical Officer of the Privy Council, 1863, p. 240.
[19] Bailey and Culley, p. 167.
[20] E.g.:

    1820: wheat, 67s 10d; barley, 33s 10d; oats, 24s 2d per quarter.
    1837: wheat, 55s 10d; barley, 30s 4d; oats, 23s 1d per quarter.
    1864: wheat, 40s 2d; barley, 29s 11d; oats, 20s 1d per quarter.
    1873: wheat, 58s 8d; barley, 40s 5d; oats, 25s 5d per quarter.

[21] *The bread of our forefathers*, 1928, p. 17.
[22] Bailey, pp. 165–6.
[23] Poor Law Commission report, 1834, section B1–2–98b.
[24] Bailey and Culley, p. 175.
[25] 1864 report (appendix), p. 15.
[26] Pococke, p. 211.
[27] Dickinson, p. 59.
[28] Mackenzie, 1825, vol. 1, p. 207.
[29] 1842 report (Mitchell), p. 766.
[30] Ibid, p. 769.
[31] Ibid, p. 761.
[32] Ibid, p. 762.
[33] A. K. H. Jenkin, *The Cornish miner*, 1927, pp. 249–52.
[34] 1864 report (appendix), p. 15.
[35] Granger, p. 21.
[36] Newcastle and Carlisle Railway Bill evidence, 1829.
[37] J. Bell, *Greenwich Hospital collections*, MS in Newcastle City Library.

[38] J. Housman, *Topographical description of Cumberland*, 1800, p. 43.

[39] Broadside posters advertising Haltwhistle market in Northumberland Record Office.

[40] T. F. Bulmer, *History of Northumberland* (Hexham division), 1886, p. 381.

[41] J. Nicholson, 'Wolsingham memoranda'. *Proceedings of the Newcastle Society of Antiquaries*, third series, vol. 2, p. 208.

[42] Bailey, p. 281.

[43] Newcastle and Carlisle Railway Bill evidence, 1829.

[44] Nicholson, p. 208.

[45] Featherston, p. 17.

[46] Sopwith, 1846, p. 13.

[47] 1842 report (Mitchell), p. 749; Raistrick, 1938, p. 40.

[48] B/B 95.

[49] B/B 50, 23 November 1795.

[50] B/B 50, 11 December 1799.

[51] Adm. 66–100, 7 February 1800.

[52] Adm. 66–99, 16 June 1796.

[53] Adm. 66–100, 21 February 1800.

[54] L.L.C. 15, 20 June 1800.

[55] Raistrick, 1938, pp. 35–6.

[56] B/B 51.

[57] Raistrick, 1938, p. 37.

[58] L.L.C. 23, 29 November 1838.

[59] Adm. 66–100, 21 February 1800.

[60] Raistrick, 1938, p. 37.

[61] Adm. 66–98, 22 February 1784.

[62] L.L.C. 16a, 21 April 1810.

[63] Hexham manorial papers.

IX

Population, migration and the poor

Population levels followed the expansion or contraction of the industry, as the price of lead rose or fell, and as new veins were discovered or old ones exhausted.

Population trends

There are two types of statistical evidence about population in the region. First, the decennial census reports cover the years from 1801 onwards (see Table 4). Second, some employment figures exist, and

TABLE 4

DECENNIAL CENSUS POPULATION FIGURES

Date	Alston	Middleton	Stanhope	Allendale	Hunstanworth
1801	4,746	1,383	5,155	3,519	215
1811	5,079	2,218	6,376	3,884	386
1821	5,699	2,866	7,341	4,629	411
1831	6,858	3,714	9,541	5,540	511
1841	6,062	3,787	7,063	5,729	567
1851	6,816	3,972	8,882	6,383	615
1861	6,404	4,557	9,654	6,401	778
1871	5,680	4,579	10,330	5,397	704
1881	4,621	4,412	8,793	4,030	502
1891	3,384	3,804	8,031	3,009	271
1901	3,134	3,574	–	2,763	220

Note: the boundaries of Stanhope parish were changed several times in the nineteenth century. The figures for 1901 are omitted, as the change was so great. The 1851 figure for Alston is swelled by navvies building the railway.

these have considerable significance in an area so dependent upon one industry. The Greenwich Hospital recorded quarterly the numbers of miners, labourers and washers working on Alston Moor from 1738 to 1767 and from 1818 to 1844 (see Appendix 6). The London Lead Company's records contain irregular employment figures for its mines on Alston Moor and in Teesdale and Weardale from 1800 to 1820. The Blackett/Beaumonts did not specifically record employment figures but it is possible to calculate from the mine accounts a rather rough indication of the numbers of persons employed at different times in Allendale. The 1842 and 1864 commissions' reports contain isolated figures relating to different mines and companies.

What were the economic factors that governed employment in the lead mines? First, there was the degree of exploitation and the productivity of a particular mining field. Alston Moor was heavily exploited in the eighteenth century; by the mid-nineteenth, exhaustion was not far off and population declined with employment. Teesdale, on the other hand, was little exploited in the eighteenth century and its population increased vastly in the nineteenth. The second factor was the state of the lead market. Booms and depressions affected the whole region. Final collapse in the 1870's and 1880's led to the permanent migration of a large part of the population. Lastly, there were seasonal variations in the numbers of workers employed which had no effect on population.

The population history of the region in the eighteenth century is obscure. It would appear that population increased, at least in the areas where the mines were fully exploited. Baptisms in Alston (excluding the chapelry of Garrigill) and Allendale increased greatly. There were more than five times as many children baptised in Alston in the five years 1800 to 1804 as there had been in 1720 to 1724, and about three times as many in Allendale.[1] These relative increases coincided with the expansion of employment in the lead industry.

When the Greenwich Hospital took over the confiscated Derwentwater estates in 1735 the Alston Moor mines had been little worked for many years. In 1738 only 300 miners were employed on the moor; in 1744, at the nadir of a slump in lead prices, there were less than 200. In 1766 over 1,400 miners were employed. Alston had 386 houses in 1750; in 1781 there were 865.[2] A Greenwich Hospital report of 1774 noted that the population of the town of Alston had 'greatly increased of late owing to the flourishing state of the mines'.[3] There seems thus to have been a substantial increase in the population of Alston in the period 1740–80. After 1780 until the end of the century the increase apparently slowed down. The 1790's were a period of depression. Employment figures have not survived, but in 1794 it was estimated that the mines employed 'near 1100 men', i.e. fewer than in 1766.[4] In 1802 the London Lead Company's chief agent recorded that the 612 men employed by the Company on Alston Moor and in Teesdale amounted to 'more than have been employed for the last 10 years.'

Far less statistical material has survived for the other mining areas in the eighteenth century. The Allendale account books, however, do show that the overall trend was one of expansion. There were about twice as many men employed at Allenheads in 1790 as there had been in 1760, and nearly three times as many at Coalcleugh. The number of baptisms went up from 301 in the period 1760–64 to 548 in 1800–04. Teesdale was comparatively little worked in the eighteenth century. The population of Middleton parish in 1801 was only 1,383, compared with 4,746 in Alston, a parish of comparable size. The Weardale and Derwentdale mines were worked more extensively, but there are no reliable data about population changes before 1801. John Wesley, visiting Weardale in 1772, however, made the revealing comment that the sides of the dale 'were sprinkled over with innumerable little houses, three in four of which (if not nine in ten) are sprung up since the Methodists came hither' in the 1740's.[5]

The eighteenth-century evidence suggests that the population of Alston probably at least doubled between 1740 and 1800. In Allendale the population appeared to increase considerably between 1760 and the end of the century. In neither case is there any evidence in parish registers or elsewhere to suggest large-scale immigration in the later eighteenth century. A few skilled miners from other parts of Britain did come, but not, it seems, in large numbers. The mining area of Teesdale (Middleton parish)—comparatively undeveloped in the eighteenth century—nearly trebled in population between 1801 and 1831; again, there was no significant immigration from outside the region, though there was some movement into Teesdale from the other lead mining parishes. In the region as a whole (i.e. the parishes listed in Table 4), population increased by 75·5 per cent between 1801 and 1831 (compared with an increase of 56·3 per cent over the whole of England and Wales). The inference is that in these years and in the later eighteenth century, population increases were due largely to natural causes, i.e. an increase in the birth and survival rates, a reduction in the death rate, or a combination of the two. 'It seems clear that the rise in the birth rate was closely connected with the process of industrialisation' in eighteenth-century England.[6] In the lead mining region an expanding economy and an increasing population were closely linked; distinguishing the cart from the horse is as difficult as in England as a whole.

With the institution of the decennial census at the beginning of the next century the population history of the lead mining region becomes less obscure. In every parish, population increased substantially in the thirty years to 1831. The population of Stanhope increased by 4,386 (85·1 per cent); of Middleton by 2,331 (168·5 per cent); of Alston by 2,112 (44·5 per cent); of Allendale by 2,021 (57·4 per cent); of Hunstanworth by 296 (137·7 per cent). Immigration from outside the region was not a significant factor in these increases.

The rise was halted by the depression of the early years of the decade 1831–41. The populations of Allendale, Middleton and Hunstanworth increased very slightly, that of Alston fell slightly, and the population of Stanhope decreased substantially—by nearly 2,500 inhabitants. In the second quarter of 1830, 1,499 miners were employed on Alston Moor, in the same quarter of 1833 only 863. The Poor Law Commission, collecting evidence in Alston in 1832, was told that some 2,000 inhabitants had left the parish since the 1831 census.[7] Some 500 'heads of families' left Stanhope parish almost immediately after the census had been taken.[8] By 1841 the depression was over and many of the migrants had returned.

Alston parish reached a peak in 1831, although twenty years later its population was only 42 fewer. The populations of Allendale and Hunstanworth began to decline after 1861, and those of Stanhope and Middleton after 1871. In 1891 Alston had 3,020 inhabitants (47·1 per cent) fewer than in 1861; Allendale had 3,392 (52·7 per cent) fewer; Hunstanworth 507 (65·2 per cent) fewer; Middleton 753 (16·6 per cent) fewer; Stanhope 1,623 (16·9 per cent) fewer. This tremendous population loss over thirty years was the direct result of the introduction of cheaper supplies of lead from abroad. The Beaumonts abandoned mining in Weardale in 1883, and in Allendale in 1884. The London Lead Company abandoned its Alston Moor leases in 1883.

The census figures illustrate the general rise and decline of population during the nineteenth century, and the direct connection between population and the fortunes of the trade. They also show interesting differences between the dales. Why did the population of Middleton increase by 168·5 per cent and that of Hunstanworth by 137·7 per cent between 1801 and 1831, whilst that of Alston increased by only 44·6 per cent? Why did Stanhope and Middleton lose a far smaller proportion of their inhabitants between 1861 and 1891 than the other lead mining parishes? There are in fact two factors at work here: (1) the degree of exploitation and the productivity of particular mining fields and (2) the growth of other industries in certain dales.

All individual mines went through periods of expansion and regression. Coalcleugh, one of the two mining centres in Allendale, was very much smaller than Allenheads for most of the eighteenth century. Yet in the 1790's, and again around 1812, more men were employed there than at Allenheads. After 1813 the mine entered a period of semi-exhaustion which extended to the 1840's. Employment figures remained low even when the lead trade was booming. In 1816 the Coalcleugh agent reported that 'The mines at Coalcleugh are extremely poor. . . . The earnings [of the miners] are now and will be. . . very low, and without that we meet with something shortly more productive than we hope, it will be impossible to continue the number of Workmen that are now employed.'[9] In 1820 'A part of the Workmen would have been obliged to have left the field had we not been able to give them employment at Allenheads Leadmines.'[10] Fifty years later Sopwith noted in his diary that 'These mines, worked for centuries, were thought to be nearly exhausted . . . but since that time . . . a large quantity of ore has been raised.' In the early 1850's Coalcleugh was again employing more men than Allenheads. But by 1869 the Coalcleugh pay was 'by the increasing poverty of the mines reduced to about a tythe of its amount ten years ago'.[11]

The London Lead Company abandoned the Derwent field as exhausted in 1806. The mines were re-opened by another concern a year or two after. In 1811 the Lead Company's chief agent reported somewhat disconsolately that 'The Mining Countries are getting into great distress and the people are flying in all directions for employment and none to assist them but Frederick Hall who employs all that go into Derwent'.[12] In 1842 a Derwent miner recollected that immediately after the Lead Company left, there were only eight men working there; in 1842 there were some 440 pickmen and labourers.[13]

The most striking example of the exhaustion of a whole mining area is the history of Alston Moor in the second half of the nineteenth century. Population increase in the first thirty years of the century was relatively less than in the other mining areas and the final decline began earlier. In 1857 the London Lead Company's chief agent told the Select Committee on the Rating of Mines that on Alston Moor 'the body of the ore is already gone; there have been in past ages most extensive and spirited workings there, and the cream of the ore . . . is gone from the district, and we are now only left to pick the leavings of others'.[14]

Teesdale, little worked until the nineteenth century, was so rich as to be worth continuing after the collapse of lead prices. The London Lead Company still employed 220 pickmen there when it disposed of its Alston Moor mines in 1884.[15] In Weardale, too, lead mining persisted, owing to the discovery of the extraordinary rich Boltsburn 'flats' in 1892.

These lately discovered riches were not, however, the major reasons why the population in Stanhope and Middleton fell less than in the neighbouring parishes at the end of the nineteenth century. The remaining workers employed in lead mining and their families were a comparatively insignificant part of the population. In Weardale, for example, the Beaumonts employed 1,003 men and boys in 1842,[16] 1,163 in 1864,[17] 720 in 1873[18] and 230 in 1882.[19] The great Burtree Pasture mine had employed 475 men and boys in 1864;[20] in 1880 those employed had shrunk to 'about 40'.[21] The most important reason why this crippling fall in lead employment is not reflected in the census figures to the same extent as in the other mining parishes was the growth of other industries.

In Weardale there had been a boom in iron mining in the 1850's. Iron was first discovered on the borders of Stanhope and Wolsingham parishes in 1845; by 1856 the Weardale Iron Company employed 1,700 men.[22] The iron deposits were almost exhausted by 1880 but extensive limestone quarrying was by then going on around Stanhope and in the eastern parts of Weardale.[23] In Middleton parish, too, limestone quarries were intensively worked by the 1880's. The chief reason, therefore, for the comparatively small population loss of these two parishes was increasing quarrying activity in their more easterly parts. In their western extremities lead mining and, in the twentieth century, fluorspar mining continued to a greater extent than in the other mining parishes, but population loss in these western areas was very much greater than in the parishes as a whole.

Migration

Migration has been frequently mentioned in this account but the evidence relating to it has so far been implied rather than documented.

The uncertain fortunes of the industry and the absence of alternative work gave the lead miner no security of employment; movement within or from the region was something that could be forced upon him at any time.

Immigration into the region. There is no evidence of *large-scale* migration into the lead mining region during the late eighteenth and nineteenth centuries.[24] What did occur, apart from the odd haphazard migrant, was an incursion of a group of skilled men from other metalliferous mining areas. About 1720 a 'Company of Mining Adventurers' came from Wales to Alston to prospect for copper ore. 'According to tradition,' they mistook a bed of sulphur for copper ore and 'erected a building for smelting it. . . but of course were disappointed of getting Copper.'[25] Some thirteen families of Derbyshire miners came to work Langdon Beck mine in upper Teesdale in 1758; some subsequently returned to their native country but others stayed and intermarried with the natives.[26] In 1796 a Cornish miner by the name of Richard Trathan came with his sons to Alston Moor, introducing a number of technical innovations in the dressing of ore. One of his sons and many of his grandchildren were living at Nenthead in 1851. Cornish and Welsh miners were introduced to the Derwent mines in the 1850's on a scale sufficiently large for their presence to be commented on in a note to the printed schedules of the 1861 census.

These immigrants, however, came in relatively tiny numbers. Negative evidence suggests that even in the later eighteenth century, when the industry was expanding fast, there was no large scale incursion. There is nothing in parish registers or mining company records to suggest it. In the nineteenth century only in Derwent was there significant immigration from *outside* the region. In 1842, of 150 'boys and young persons' employed in the Beaumont Weardale mines, only one was born outside the parish of Stanhope. In Allendale, out of 252 boys, twelve came from outside the parish—but only from Alston.[27] Of the inhabitants of Nenthead in 1851, only 58, or 6·6 per cent, were born outside the lead mining region. More than half these 58 were members of families in no way engaged in lead mining—professional and landowning people, shopkeepers, and lodging-house keepers and their lodgers. Many of the remaining 27 were children of lead miners born while their parents were temporarily absent from the area during the depression of the 1830's, or were from the other metalliferous mining areas of Britain.[28] The lead mining region had little appeal for immigrants without previous mining experience, and was too isolated for many people to drift to it.

Migration within and from the region. On the indigenous inhabitants the region had an unusually powerful hold. The Poor Law Commission was told in 1834 that

The leadminers in this district, like other mountaineers, cherish extraordinary attachment to the place of their birth, occupations, and habits ... [The depression in the lead trade] throwing up on the surface of the soil a population which had previously drawn sustenance from its bowels, it would seem must end in a state of things unparalleled in wretchedness, amongst a people obstinately clinging to their native place, and in a tract of country quite unable to feed its own inhabitants.[29]

In 1842 Commissioner Mitchell wrote of Weardale:

Altogether the natives of the dale grow up with an attachment to their native land and their own people which nothing can overcome. Hence it is that, although by removing only 20 miles down into the coal country a young man might nearly double his income, and have the prospect of adding many years of health and strength to his life, he cannot remove. He clings to his beloved dale, and follows an occupation which in most instances allows but a short life, the last years of which are spent in sickness and sorrow.[30]

Why then did miners migrate? The most important cause was prolonged unemployment due to a depression in the lead trade, or exhaustion of a particular mine or field. Unemployment was not uncommon; the 1834 Poor Law Commission was told that in normal times, in Weardale, 'from the peculiarity of this kind of unemployment, there may be generally one tenth of them out of work.'[31] When lead prices were low, one of the ways in which the mining companies habitually economised was by cutting down their labour force. For example, in 1808 the London Lead Company's chief agent reported:

I have reduced the hands on the Dead Work List from 162 ... to 104 ... and have only kept in sight the preservation of the leases [which required a certain minimum number of men to be working at a given mine]. Under the circumstances, I flatter myself the Business at large will do well; although Poverty begins to stare the Country very much in the face, and the general murmur of the labouring Class renders the lives of public men very troublesome.[32]

In 1815 the company's Teesdale agent, Robert Stagg, wrote that there was 'a Reduction of eleven pickmen this Quarter which are all I could dismiss with propriety . . . I expect to dismiss a similar number next Quarter, which will reduce the Workmen as low as I conceive to be advantageous.'[33]

Similarly men had to be dismissed when a mine was (or appeared to be) exhausted. In Weardale in 1822:

A considerable part of the Mines which have for a series of years past been a great support to the produce of Ore, and given employment to a great number of Men, are very much worked out. . . . These circumstances have caused a greater number of Men at this time to be out of employment than usual; and except some new trials should prove more productive than there is any appearance of at present, it will not be in my power to give employment judiciously to the increasing population of this district.[34]

In the 1860's Sopwith's diary records the difficulties he had when 'the exhausted state of part of the Allenheads mines renders the discontinuance of some of the several men . . . a matter of hard necessity . . . The consideration of the principle of selection for workmen to be retained or discharged . . . is a duty which causes me no little anxiety.'[35]

Miners sometimes relinquished their jobs voluntarily—when the bargain prices offered were so low that they would not cover their subsistence. In 1831, for example, at Sedlin mine in Weardale, it was 'so extremely poor that not one half of the usual number of Men have taken bargains. Since last quarter 20 Men have left and gone to the Coal works, etc., and 20 more have refused taking bargains at the price offered, and who are now in want of employment.'[36] This was at a time when monthly subsistence payments had probably been reduced owing to the severity of the depression; at other times miners often preferred to work on, in full awareness that they would not make enough to cover their subsistence debt: 'several of the Bargains at different Mines I fear are taken with a View of having their advance Money continued.'[37]

Thus being out of work was no rare event for the employees of the two largest concerns. It was probably even more familiar to the employees of the smaller ones. An Alston miner told Dr Mitchell in 1842 that 'Men working under the great companies have their work more regular'.[38] Migration, however, did not automatically follow, despite the almost complete absence of alternative employment.

Three things cushioned the effect of unemployment over short periods. In the first place, everyone knew how variable were the factors governing the level of employment. Lead prices could rise almost overnight, and labour would once again be in demand. A new vein or a flat might be encountered, turning a poor mine into a rich one. In view of these circumstances unemployed men were prepared to remain in the region for as long as possible. In the second place, many of them belonged to families that owned smallholdings, from which a bare sustenance could be eked. Lastly, at least until the 1830's, there was the parish, and perhaps enough outdoor relief to make it possible to exist for a time without work.

Thus unemployment had to be prolonged to force miners to leave. It was the length of the depression that distinguished the early 1830's from the period around 1816, when lead prices tumbled almost as low. There was great distress in 1816 and some miners did migrate, but the later slump forced far more men and their families to leave. As we shall see, however, many returned when the worst was over.

If unemployment was the most important cause of migration, there were sometimes other factors. The specialist workers—particularly the smelters and refiners—could often get better wages by moving to another mining company. In 1851 only thirteen of the 133 active lead miners in Nenthead had not been born in the village, but sixteen of the 31 smelters were not natives. In addition, miners were sometimes dismissed for disciplinary offences such as going on strike, drunkenness or having an illegitimate child. Dismissal 'was considered the severest of punishments', the men—as Stagg, the London Lead Company's chief agent expressed it—'haunting the place like ghosts for months afterwards.'[39]

From about 1865 onwards there was something of a 'wind of change' blowing through the lead mining region. Unfortunately this closing period of its history is poorly documented, but a few scraps of evidence suggest that the region no longer exerted so strong a hold over its inhabitants, and that the financial and other attractions of the outside world were beginning to be felt. In 1866 Sopwith commented in his diary upon what he considered the Beaumont land agent's mistaken policy of raising rents:

The old and infirm must linger. The young and active depart. But along with this is not increase of rent suggestive of the desire for increased wage, and this wish may gather strength as railway works afford demand for labour when they are in construction, and when finished will give facilities for movement and for interchange of opinions respecting wages such as have not hitherto been known in these secluded dales.[40]

This change in the miners' attitude and their greater willingness to leave the district before economic necessity forced them to is illustrated by the surviving report books of the Derwent Mining Company, covering the period 1872–79. The Derwent mines were particularly close to the coal pits and ironworks, so they may not be typical of the region as a whole. Here are a selection of quotations from reports in the period 1872–73:

... At the setting on Saturday there were five refusals—Some of our men have left us for other places where work is so plentiful. ... [29 January 1872] ... And altogether our weekly liftings have recently receded a good deal, not altogether because of the Mines being poorer but also because of the scarcity of miners. ... We want *more Men* ... [21 June 1872] ... The Setting on Saturday was rather thinly attended, some half-Dozen bargains refused, some of which have since been set. Such great inducements are offered at the Coal Mines about Consett and further east, and also down the Wear that a few of our Men, the young ones especially, some of whose earnings have been between 20/– and 30/– a week the last month, keep drifting away. It puzzles us what to do to keep them ... [30 September 1872] ... Since the Setting we have lost 4 men, 2 of whom had been neglecting their work very badly, and when reproved for their negligence, they gave up their place of work, and left for the coal mines ... [13 January 1873]

A very different feeling towards their native place and occupation is shown by these miners compared with their dismissed predecessors of the 1830's 'haunting the place like ghosts for months afterwards'.

When the miners migrated, where did they go? It often happened that when the cause of unemployment was local exhaustion rather than general depression, they did not leave the region but went to work at a different mine. The lodging shop system permitted them to work a long way from home, so there was frequently no need to move house. A note in a London Lead Company report book for 1810 suggests that it was regular practice at that time for Alston Moor miners to prefer working for small companies in times of boom, and return to the company when times were hard. 'Many parts of the Mines are much poorer this quarter than at Michaelmas last, though a great number more Hands are employ'd, by reason of the Mines in the surrounding Country being so poor, of course the Miners cannot get employment elsewhere.'[41]

There are frequent notes in the Beaumont records of men transferring from a poor mine to a rich one. Coalcleugh miners went to Allenheads in the 1820's, and in Weardale miners went wherever they could get work. In 1812 'Sedlin mine . . . may be compared to a Hospital, taking all those that hath not employment elsewhere.'[42] In 1860 unemployed men were listed in the Weardale bargain book, to be put into any mine where there were vacancies. Similarly, London Lead Company employees moved around on Alston Moor and in Teesdale.

Transferring from one company to another was scarcely less common. The movement to the Derwent mines in 1811 has already been mentioned. A year before, a number of Nenthead men applied for work at Coalcleugh, as they were dissatisfied with 'the irregular behaviour of Mr. Dodd [the Lead Company's agent] of late'. They were employed 'to point out the different strings etc. about the end of the Boundary' between Coalcleugh and Nenthead.[43] Alston miners were brought in to break the 1849 Allenheads strike. In 1859 'troops of young men and boys' from Allendale went to work in Settlingstones mine in the Tyne valley every week, returning at the weekends.[44] In 1864 miners who lived in Allendale were working in Derwent.[45]

When miners left the region, they generally took their families with them. The accounts of the large-scale migration in the 1830's frequently use the word *families* rather than miners to describe those leaving. R. W. Bainbridge, the London Lead Company's chief agent, was questioned on this very point by the 1857 Select Committee on the Rating of Mines. He replied that if all the mines closed down

there would be no miners resident in the district: they would have emigrated and found employment elsewhere. . . . The bulk of the able-bodied would take their disabled away with them. There are such family attachments . . . that very few able-bodied parties would leave their homes, and leave the infirm of their families behind them.[46]

The normal destination of the miners was the coal area of the north-east coast. Eden specified 'the coal-mines near Newcastle, Sunderland, etc.' as being where migrant Weardale miners had gone to work in the 1790's.[47] Of ten miners who had left Coalcleugh in 1824 owing lent money, five had gone to other lead mining areas, three to the collieries and two to America.[48] In the 1830's the vast majority of migrating miners went to the north-eastern coalfield, and others went to the mines around Whitehaven, but some went to Spanish America,[7] and a fund was raised on Alston Moor 'to assist such poor persons as may be desirous to emigrate to Canada'.[49] In 1832, 124 people left Alston for Canada. There is little evidence about the destinations **of** the mass of immigrants who left the region between 1860 and the end of the century. The probability is that the still expanding north-eastern and Cumberland coalfields absorbed most of them, but many left England altogether. A 'memo' in the Weardale bargain book for 1860 gave 'a list of workmen who have left the works' the previous quarter. Of the twenty-eight who had gone, fourteen went to Australia, one to Columbia, one to work on the Rookhope railway, and one to work in the ironstone works. The destinations of the remainder were not specified.[50]

Migration between 1830 and 1842. For the period 1828–42 an unusually large amount of evidence regarding migration has survived. The severity of the depression at this time occasioned frequent references to the miners' distress in the London Lead Company and Beaumont records; the Poor Law Commission, reporting in 1834, gathered evidence at the height of the crisis; the commission of 1842 found plenty of memories of the period among the miners questioned; and, most usefully, the coal viewer Matthias Dunn recorded in his diary many details about the employment of lead miners in the coal mines during the period 1831–34.

The early 1830's saw the price of lead sink to a level it had not reached for fifty years. In the mining region the effect on employment was disastrous. Thousands of men were thrown out of work. In the minute book of the London Lead Company for 1830 is a report of a sub-committee deputed to treat with the Greenwich Hospital for a reduction in the dues payable 'in order to enable the Company to employ as many Labourers as possible in the present distressed state of the Mining Population with the least injury to the Company'. The Greenwich Hospital Commissioners, however, would not agree

to any alteration that might tend to increase the raising of Ore considering it a better Alternative that the Men now employed in mining should be reduced and obliged to find other employment which they thought would be practicable in making Roads and other public Works. . . . From the tenor of the Conversation that passed, we are of opinion that the Company would be completely exonerated from any blame by taking such measures in the future Workings of the Mines in Alston Moor as may appear most beneficial to their Interest in accordance with the general Terms of their present Leases.[51]

The Lead Company dismissed hundreds of men; smaller companies on Alston Moor virtually ceased operations. The Beaumont reports of

o

the period chronicle for every mine 'a number of men out of employment'. The Greenwich Hospital's suggestion that the unemployed
should turn to public works shows one way in which some miners
could find alternative employment—for a while at least. One of the
receivers of the Greenwich Hospital wrote, on 4 December 1830,

we propose the continuing of the Nent Force Level on a contracted scale
and road improvements. These afford the best field for the employment of
capital by those owners of property within the district who wish to combine
a certain public benefit with a reasonable prospect of advantage to themselves.[52]

The sentiment was shared by landowners in other districts. In Weardale the Poor Law Commission was told that 'Many landholders, to
furnish employment, have been extensively engaged in improving
their land'.[53]

These 'public works', however, could not absorb everyone, and a
mass migration started towards the end of 1831. In Alston nearly
2,000 inhabitants went out of the parish—'the population are obliged
to disperse in all directions', as the Poor Law Commission was told.[54]
In Weardale 'some 500 heads of families' left. 'The firm refusal of the
parishes to make up insufficient wages forced numbers of the lead
miners to transplant themselves to the collieries.' Conditions in the
lead mining parishes were, however, mitigated 'by the fortunate
coincidence of the grand strike in the collieries.'[29]

The strikes of 1831 and 1832 were the culmination of many years
of discontent among the colliers on the Tyne and Wear. From April
1831 to September 1832 there was a series of bitter strikes which
ended in the almost complete victory of the coal owners and in the
disruption of the colliers' union. The coal owners found in the unemployed lead miners a convenient source of 'blackleg' labour, and it was
largely the ease with which they could draw on it that eventually
defeated the striking colliers utterly. Lead miners were brought in in
1831, but few details survive. At Waldridge colliery on 24 December
a body of 'upwards of 1000 pitmen riotously assembled'. There were
thirty or forty lead miners down the pit, and the striking colliers
'stopped the engine for pumping out the water, and threw tubs, corves,
etc. down the shaft, until dispersed by a body of the military'.[55]

From the end of 1831 the diary of Matthias Dunn, coal viewer of
Hetton colliery, contains a series of notes recording how lead miners
were attracted to the collieries, the ejection of colliers from their

homes in favour of the lead miners, and the unemployment of many of the 'stranger colliers' once the strike was over:

At Coxlodge Colliery endeavouring to procure Men for Hetton as they are discharging the Colliers from their Houses to prepare for Lead Miners ... [16 January 1832]. Lead Miners now determined upon ... [19 March] W. Robson went off to the Lead Mines in search of Men, provided ours will not be bound ... [27 March]. Meeting of the Hetton Committee at New-castle to consider matters generally—1. To the getting down of London police. 2. Sending up to the Lead Mines for additional hands ... [10 April]. I started for Alston Moor—where I appointed an Agent and staid all night ... [11 April]. Crossed the Country to Middleton in Teesdale where I saw a number of Men willing to come, but who wish some time for consideration ... [12 April]. Turned out about 20 families chiefly marked men. ... No violence offered—women even quiet and submissive ... [21 April]. Sep. Redhead started for Weardale and George Lisle and I started for Teesdale and Arkendale in search of Miners. Turning out determined to be carried on successively at 20 and 30 per day, but difficulty is experienced in procuring assistance ... [23 April]. At Middleton all day collected good many Miners ... [26 April]. A little Coal work beginning at Isabella Pit—sundry Lead Miners arriving who require to be escorted through the Pitmen by Military ... [3 May]. Got the help of the police and ejected 80 families at No. Hetton. Lead miners gradually arriving. Isabella Pit now getting 20 per day ... [7 May]. Remainder of people ejected from their Houses, Hand Bills printed for Lead miners and every exertion now to be tried to supply the place of these foolish men—Events will prove that the Men of this Colliery will be more humiliated by the introduction of Strangers than any other as the work is so extremely simple and Comfortable. Measures are now taking for distributing Hand Bills in the Mining District for procuring men with all convenient despatch ... [8 May]. A good many Miners arrived with their Furniture and peacably enough received ... [21 May]. George Pit started also Isabella having now more men than she can hold. Pitmen still stupid as ever and seem determined to stick out to the last notwithstanding the obvious Storm that is gathering around them ... [14 May]. Are now endeavouring to procure Men from Yorkshire, Wales, Lincolnshire, Derbyshire, Cheshire and Cornwall, etc. ... The Pitmen stand out firmly, about 40 to 50 persons only having come out of the Union ... [31 May]. Have now about 1200 Strangers, Men and Boys ... [16 June]. Number now upwards of 1700 ... [26 June]. Union all but broken up ... [18 September].

These entries have been quoted verbatim as they convey something of the drama of the situation. In 1834 there was a mild depression in the coal trade, and Dunn notes, 'Great many stranger colliers going away in Consequence of the reduction of prices ... They are incited by the old pitmen evidently under the notion of a future stick' [14 April 1834].

The movement of miners from the lead dales in 1831–32 was not, then, due solely to the depression and resulting unemployment. It was also due to large-scale attempts (Hetton was only *one* strike-bound colliery) to attract men to the coal mines to replace the striking colliers. Once the strike was over, there was a superabundance of labour on the coalfield. The incoming lead miners were regarded with hatred by the colliers who had been expelled from their homes. Not surprisingly, there was a movement back to the dales by some of the migrants. The 1834 Poor Law Commission was told of fears that 'the migratory swarms' would return 'partly from failure of employment, and partly the . . . attraction of the *natale solum*. Many may thus be expected to fall back on their original settlements, as from the terms of their hiring they cannot, for the most part, have gained settlements elsewhere'.[29] Fortunately, the price of lead started to rise again in 1833.

The 1842 commission recorded many examples of lead miners who had left the area and later returned to it in the 1830's. Here is a Weardale miner telling the commissioner why he went—and why he returned:

I have known when a young man working in the mines has got only 1/– a day; that was 10 years ago. Many went down to the coal district and got 3/– a day; some of them remain there, and some have returned. I went down myself, but my parents entreated me to return, and my heart softened and I came back. My brother was getting only 7/6 a week about three years last spring, and went to the coal pits and would average 50/– a fortnight; we persuaded him to return.[56]

To sum up, then: until the middle of the nineteenth century, before railways came to the region, emigration—unless forced by economic crisis—was slight. The industrial attractions were insufficient to draw any large-scale movement *into* the area. When the miners did migrate, they went mainly to the nearest large centres of population—most to the collieries and ironworks of Northumberland and Durham, some to those of west Cumberland. Those who had families took them with them. The existing evidence indicates that most continued as *miners* and did not become, say, shipbuilders. After the middle of the century there was a greater readiness to move, and more emigration from Britain altogether.

Poor relief

The amount of poor relief given to unemployed but able-bodied lead miners obviously affected migration considerably. Unfortunately the evidence here is particularly sparse. Not a single overseers' record book survives for any of the lead mining parishes, so the workings of the old poor law during the eighteenth century are virtually unknown. Fortunately, there is some detail in the 1834 Poor Law Commission's report concerning the parishes of Alston, Middleton and Stanhope, and the medical officer of health's records in the Public Record Office cast some light on the working of the new poor law after 1834.

The parish was not the only source of poor relief. The larger mining companies supplied their miners with subsidised bread in times of high prices, and tried to keep the mines going even when the price of lead made it unprofitable in the short term. The reasons were not purely humanitarian; dead work could be done more cheaply at such times, and a labour force had to be kept together for when the price rose again. The London Lead Company sometimes subscribed towards the relief of the poor in Alston or Middleton parishes,[57] prompted by the fact that until 1874 lead mines were not chargeable for poor rates—much to the indignation of the landed proprietors of the various parishes, who frequently petitioned the House of Commons to rectify this state of affairs. Those mining concerns which were also landowners had, therefore, an additional reason for seeing that mining did not collapse altogether in bad times, as otherwise their pockets would suffer through increased rates. This applied particularly to the Blackett/Beaumonts in Allendale and the Greenwich Hospital on Alston Moor. The Greenwich Hospital report of 1821 summed up the position nicely with a neat mixture of self-interest and humanitarianism. The report argued that the Hospital should not levy so high a tribute from the mining companies as to cause any redundancies among the workmen:

They would ultimately be compelled to emigrate or fall upon the Parish, and the Commissioners and Governors of Greenwich Hospital would thus become answerable for an alarming amount of human Misery, which they have the power to prevent; besides thereby bringing on the Institution the entire loss of its Revenue from the Mines, and a heavy burthen upon the Parish towards which they must ultimately contribute the principle portion.

Under the old poor law, parochial relief was given to widows, orphans and disabled men. What is not clear is to what extent it was given to unemployed able-bodied men. The fractional evidence that exists before the 1830's shows that most parishes did give some aid, though upon what principle is not known. A petition to the justices of the peace for south Northumberland from Allendale parish in 1711 stated that the laying in of the mines 'will increase the number of the poor', thus implying that assistance was given to the able-bodied at that date. Jumping a century to the period 1810–18, plenty of evidence exists in the records of both the Blackett/Beaumonts and the London Lead Company to show that unemployed men regularly applied for parochial assistance. At the Lead Company's mines in Weardale in 1816 the agent advised unemployed men 'to apply to their respective Parishes'.[58] A petition to the House of Commons from Alston parish in 1817 stated that owing to the very low price of lead 'a great number of farmers of small tenements, who heretofore contributed to the parish fund, are fast increasing the list of paupers, and instead of paying to maintain others, apply themselves for parish relief.'[59] Again in Weardale, in 1818, the miners still at work were receiving relief because their wages were so low: 'Great numbers of us are not making our Subsistence Money, & have been under the necessity of seeking relief, upwards of 400 of us (including their Families) are on the Parish' (see Appendix 4).

In the early 1830's the principles on which relief was granted were different in the several parishes. In Alston, relief was given 'after an enquiry into the gains of the family, if it does not amount to 1/3d per head weekly, to make it up to that sum'. At that time there were 216 *families* receiving such relief.[60] In Stanhope, outdoor relief was not given on this Speenhamland system. It had been proposed that 'the deficiency in the men's wages should be made up by allowances from the parish', but 'this attempt was frustrated by the firmness of a resident magistrate, the present member for Gateshead'.[61] Relief was given only through the workhouse. Outdoor relief was received by 288 *individuals*, 'two thirds of them widows and their children; the other third infirm Men and Orphan children'.[62] In Middleton, too, the same principle was apparently followed; outdoor relief should not be given to able-bodied men.[29]

The Poor Law Commission report gives some information about how the laws relating to settlement were enforced. Many of the migrating miners were likely to return from the coal districts 'as from the terms of their living they cannot, for the most part, have gained settlements elsewhere'.[29] In Alston parish, seventy of the recipients of out relief were not resident; 'these are chiefly at Newcastle and Whitehaven, it brings less expense to the parish to pay them a small sum to remain there than to bring them back'.[49]

After the enforcement of the 1834 Poor Law Act, outdoor relief to able-bodied men ceased. In subsequent economic depressions wages were no longer made up by the parish, and men without work had no choice but to migrate if they had no other source of income.

Money was not available to make up payments to friendly societies either. In 1841 the board of guardians for the Weardale Union wrote about a miner named Colling, employed by the Lead Company, who had been

disabled from working for some time. The Lead Company has a Club or Sick Fund, for their Miners, the Members of which when of a certain standing, are entitled to receive 6/– a week during sickness. The annual subscription is £1–16–0. Colling is a Member of the Club, and has received the regular allowance from it since the commencement of sickness, and has been thereby enabled to exist without Parochial Relief. His subscription for the past year having lately become due, the non-payment of which would deprive him at once of any benefit from the Club whilst its payment would secure him 6/– a week for a year to come, an Application was made last week by Colling to the Board of Guardians to advance the ammount (£1–16–0) for him.

The Guardians did advance the money, but were told subsequently that their action was illegal. The unfortunate man had to enter the workhouse to receive any poor relief.[63]

Most of the money paid out under the new poor law went to widows, of whom, owing to the early deaths of lead miners, there were always many, often with young children. In Alston parish on 19 August 1841, for example, there were 100 widows receiving relief, of whom 88 were widows of miners.[64] Their children were working as washers as soon as they were old enough.

The change in the administration of poor relief brought about a real alteration in the lead miners' lives. After 1834 their ability to remain unemployed in the lead mining region was greatly lessened. Which was, of course, precisely the intention of the originators and administrators of the 1834 Act.

NOTES

[1] In the period 1720–24 there were 121 baptisms in Alston and 180 in Allendale; in 1740–44, 142 and 300; in 1760–64, 338 and 301; in 1780–84,

638 and 498; in 1800–04, 653 and 548. These figures are cited with all the reservations about their significance usually associated with such evidence.

[2] W. Hutchinson, *Cumberland*, vol. 1, 1794, p. 522.

[3] Adm. 79/57, p. 14.

[4] Hutchinson, vol. 1, p. 215.

[5] Wesley's *Journal*, 2 June 1772.

[6] P. Deane and W. A. Cole, *British economic growth, 1688–1959*, 1962, p. 134. The authors of this work stress the need for 'detailed local research' before any sense can truly be made of English eighteenth-century population history. The eighteenth-century evidence for the lead mining region is mainly inadequate for this purpose. The growth of population in the region is, however, strikingly similar to the contemporaneous, but of course much greater in scale, growth in the West Riding (pp. 119–20). There, too, growth was due almost exclusively to 'exceptionally high rates of natural increase' in contrast to Lancashire, which 'relied more heavily on the immigration of the surplus population from surrounding areas'. The explanation of this difference, according to Deane and Cole, was that while Lancashire was urbanised by the beginning of the nineteenth century, 'the West Riding was still for the most part a county of industrialised villages' with small attraction to immigrants.

[7] Poor Law Commission report, 1834, appendix A, p. 320A.

[8] Ibid, p. 140A.

[9] B/B 53, Christmas 1816.

[10] B/B 54, Lady Day 1820.

[11] Sopwith, *Diary*, June 1869.

[12] L.L.C. 16a, 1 November 1811.

[13] 1842 report (Mitchell), p. 758.

[14] *Rating of Mines*: select committee report, 1857, p. 28.

[15] L.L.C. 34, 17 June 1884.

[16] 1842 report (Leifchild), p. 557.

[17] 1864 report (vol. 2), p. 370.

[18] W. M. Egglestone, *The projected Weardale railway*, 1887.

[19] *Newcastle Weekly Chronicle*, March 1882.

[20] 1864 report (vol 2), p. 369.

[21] *Depressed condition of the agricultural interests*. Royal Commission report, appendix to part 1, 1881, p. 230.

[22] W. Whellan, *Durham*, 1856, pp. 333–4.

[23] Egglestone, 1887.

[24] A number of twentieth-century local historians have suggested a large-scale migration of border Scots 'Covenanters' into the region in the late seventeenth century (e.g. J. J. Graham, *Weardale past and present*, 1939, p. 54). This may have taken place, but no convincing evidence exists to support the theory. It appears to be based on the mistaken idea that because

many miners in the early eighteenth century were members of the Presbyterian Church of Scotland they were necessarily Scots (see chapter XI).

[25] G.H. (Wigan) MS.

[26] The Lord Fitzhugh (*c*. 1880).

[27] 1842 report (Mitchell), p. 724.

[28] Enumerators' books, 1851 census.

[29] Poor Law Commission report, 1834, appendix A, p. 138A.

[30] 1842 report (Mitchell), p. 722.

[31] Poor Law Commission report, 1834, section B1–1–160a.

[32] L.L.C. 16a, 23 January 1808.

[33] L.L.C. 16a, 23 March 1815.

[34] B/B 54.

[35] Sopwith, *Diary*, 8 November 1862.

[36] B/B 35, 30 September 1831.

[37] Weardale, 1812: B/B 53, 10 January 1812.

[38] 1842 report (Mitchell), p. 761.

[39] Poor Law Commission report, 1834, appendix A, p. 139A.

[40] Sopwith, *Diary*, 23 June 1866.

[41] L.L.C. 16a, 20 January 1810.

[42] B/B 53, 10 January 1812.

[43] B/B 53, 1 January 1809.

[44] White, p. 452.

[45] 1864 report (vol. 2), p. 363.

[46] *Rating of mines*: select committee report, 1857, p. 28.

[47] Eden, vol. 2, 1797, p. 169.

[48] B/B 96.

[49] L.L.C. 21, 22 March 1832.

[50] B/B 131, September quarter, 1860.

[51] L.L.C. 21, 30 December 1830.

[52] Quoted by Hughes, p. 56.

[53] Poor Law Commission report, 1834, section B1–3–160C.

[54] Ibid, section B1–1–98A.

[55] E. Mackenzie and M. Ross, *Durham*, vol. 1, 1834, p. cx.

[56] 1842 report (Mitchell) , p. 762.

[57] L.L.C. 17, 20 August 1812.

[58] L.L.C. 16a, Lady Day 1816.

[59] House of Commons *Journal*, 5 February 1817.

[60] Poor Law Commission report, 1834, appendix A, p. 320A; section B1–2–986.

[61] Ibid, pp. 139A–140A.

[62] Ibid, section B1–2–1606.

[63] Public Record Office, M.H. 12, 3333.

[64] 1842 report (Mitchell), p. 767

X

Health

Vital statistics of lead miners are available only for the middle and later nineteenth century, when they were included in one or other of the various official reports. These statistics show that the average life expectancy of an adult lead miner was less than that of a worker in almost any other trade in Britain.

An analysis of deaths in the mining parishes from 1 July 1837 to 30 June 1841 is given in the 1842 report. These figures are for adult miners only, i.e. those who were over the age of 19 at death. In Alston parish the average age of the 75 miners who died in that period was 45. In Allendale, of 79 deaths, the average age was $48\frac{10}{79}$. In Weardale, of 129 deaths, the average age was $49\frac{62}{129}$, and in Teesdale, of 57 deaths, $47\frac{15}{57}$.[1] By far the greater number of all these died from some form of respiratory disease. In 1864 a similar survey of the 72 miners who had died in the Allenheads region in the previous ten years showed that the average age of death was $44\frac{1}{2}$ years.[2] Also in 1864, an investigation was carried out at various mines into the average ages of those employed *underground*. The average age of the 339 men employed by the London Lead Company on Alston Moor was 32·86; of the 77 men at Fallowfield, 30·68; and of the 289 men in the Derwent mines, 28·12.[3] As few miners were employed underground before they were 18, these figures show that there was a surprising absence of older men.

Occupational diseases

In the annual report of the General Board of Health for 1858 there is a short statistical survey of Alston, comparing it with Reath (a Yorkshire lead mining parish), Haltwhistle (a Northumberland agricultural parish), and the city of Liverpool.[4] This survey found that while the mortality of children under five years of age from pulmonary diseases

was distinctly greater in the two lead mining parishes than in Haltwhistle, it was

immensely below that which is sustained by the infantile population of the
great unhealthy city. This advantage enjoyed by the younger inhabitants of
the lead mining districts over those of Liverpool is altogether lost in more
advanced life. The women of Liverpool perish from chest affection in a rather
larger proportion than those of Reeth; in a slightly lower proportion than the
women of Alston; but the men of Alston and Reeth die in a much larger proportion than the men of Liverpool. Thus a district remote from city influences, situated in the midst of a most salubrious district, and containing
scarcely an appreciable urban character . . . loses a larger annual proportion
of its *adult male inhabitants from diseases of the chest than the unhealthiest city
in the kingdom*. That this is due to the nature of the prevalent employment no
doubt can now be obtained. It is the injurious character of the male occupation which causes Alston, the most exclusively lead mining district in England,
to be the place in which there is *a larger proportion of widows than in any other
place in the kingdom*.

It should be said that though these Victorian medical men recognised
that the lead miners were dying from some form of respiratory disease,
medical science was not sufficiently advanced to explain either what
caused the symptoms or the precise nature of the disease. In fact the
miners were infected mainly by silicosis, one of the class now known
as the pneumoconioses, or dust diseases of the lungs. It is caused by the
inhalation over a number of years of fine particles of mineral dust,
causing minute scars (fibrosis) to form within the lungs. Once there
these scars cannot be healed. A man suffering from silicosis is very
liable to catch one of the infectious respiratory diseases, and most of the
lead miners in fact died from a form of tuberculosis.

Both the two major Government investigations of the lead mining
region took place before the discovery of the tubercle bacillus in 1882,
and before the realisation that stone dust rather than 'foul air'—gaseous
vapour from explosions, or changes of temperature or pressure—was
the only cause of the respiratory condition that led to the premature
deaths of so many miners. The disease was commonly known by
miners at the time as 'miner's complaint', and the men attributed it to
all or any of the above causes.

Miners' answers to questions about their health, as printed in the
1842 and 1864 reports, concentrate on the two most obvious symptoms of the disease—shortage of breath and 'black spit'. These symptoms first made their appearance after three or four years' work in the

mine. Wheezing and noisy breathing were commonplace. The men used picturesque terms to describe the expectoration of a bluish mucus, 'an acknowledged concomitant of a lead miner's existence'.[5]

A miner said in 1842 that he frequently spat 'black stuff, as black as the wine in your glass'.[6] Another spat 'as black as your hat'.[7] An agent told the 1864 commission that he knew of several instances 'where men have vomited up what was just like a collection of dust'.[8]

A rather harrowing description of the progress of 'miner's complaint' was given to the 1864 commission by Dr W. Ewart, the London Lead Company's resident surgeon at Middleton in Teesdale:

A healthy young man enters upon work in a lead mine; in a few years, more or less, he begins to experience some degree of difficulty of breathing—nothing to hurt him very much so he continues an efficient miner; still, however, he is 'touched in the wind'. Along with this difficulty of breathing there is an increased expectoration of mucus, often tinged, more particularly after leaving work, of a bluish black colour. The difficulty of breathing continues, and generally increases as age advances, rendered worse, perhaps, by an attack of bronchitis now and then supervening. On recovering from these attacks he again resumes work; or should it happen that he has no acute attacks, he goes on working with increased shortage of breathing, expectoration, and failing strength. His appetite for food is impaired, and what he takes for breakfast is frequently vomited as he walks to his work in a morning. He has great languor, and frequently fits of severe coughing, and evidences of imperfect oxygenation of the blood in the blueness of the lips, etc. He may have now got to 40 or 45 years of age; he is low in health and strength, and is compelled to give up work, and stay at home as a worn out miner. But even at home his health cannot be restored. He may wander about for some years, incapable of work, from difficulty of breathing and debility, in the summer, improving a little, but in the autumn, winter and spring being very liable to acute diseases of the chest producing more copious expectoration. . . . The poor worn-out miner now soon dies exhausted.[9]

The main cause of this distressing end was the dust inhaled in the course of work, though this was not recognised in the report of the 1864 commission and hence no specific recommendations were made about it.[10]

A feature peculiar to the northern Pennine lead region was the existence of the 'shop' which provided sleeping quarters at the more isolated mines. The discomforts of the 'shops' were recognised, but not their probably crucial role in spreading infectious lung diseases. When men with a predisposition to such diseases were sleeping in close contact with others already infected who were coughing and expectorating all night, it must have been difficult for anyone to escape infection.[11]

The miners accepted poor health and a shortened life span as 'a matter of course', according to a Lead Company surgeon in 1842.[12] However, a number of medical witnesses in 1864 commented that, when questioned, the miners were reluctant to admit they possessed the symptoms of the disease. One witness reported that 'Out of 100 men whom I examined ... I found that scarcely one man would complain of any expectoration, although I am aware that nearly all the men do expectorate after working a few years in the mine'.[13] Post-mortem examinations were not carried out, for moral reasons— 'there is a great objection to that sort of thing here'.[14]

When the miners felt their health was breaking down to such a degree as to make climbing ladders, drawing deads, etc., particularly exhausting, they applied for easier jobs. In the Weardale bargain book for 1860 is a note: 'Jacob Hetherington wishes to be changed his work. He has been poorly for some time past. Josh Peart is working his work.'[15] Such men were placed in 'the shallower workings, where there is better ventilation and less fatigue'.[16] Here, of course, they would be earning less. Others working in family groups were helped by their sons. Eventually badly affected men had to leave the mines, and indeed stop working altogether. Those who possessed small-holdings would retire to them until death; others would subsist on what they were given by the parish. (None of the benefit societies gave pensions to members on the grounds of old age alone until they reached the age of 70—a nearly impossible age by lead mining standards.)

The questioning in 1842 and 1864 revealed some interesting attitudes among the employers as well as among the miners. Most were reluctant to admit there was any health problem, and some went so far as to deny its existence. In 1842 the agent of the Beaumont concern 'boldly asserted that the miners lived even beyond the average duration of life'. A paper was submitted, attempting to prove that 'the miners connected with his mine who have died for the 28 years past had averaged, one with another 51 years'.[17] One commissioner, Leifchild, accepted this and printed it as an appendix to his report. The other, Mitchell, made his own statistical investigation and found that the figures the employers had submitted were calculated in such a way as to produce a misleading impression.

Even Sopwith was reluctant to admit that lead mining was directly responsible for a killing disease. He blamed bad food, insanitary houses, intemperance and stupidity on the part of the men.[18] His subordinate, agent for the Coalcleugh area, when asked if the miners were 'at all affected in their breathing', replied 'Some of them, after they arrive at about 50 years of age or so are affected a little'.[19] The Derwent Mining Company's agents followed the same line, and even produced two miners to witness that they knew of no one under 50 whose breathing had been affected by the work.[20] Only the London Lead Company's agents were completely honest about the mortality rate.

Medical services

Doctors were available in the dales even in the eighteenth century, but there was little they could do to help. In 1733 there was a surgeon living at Allenheads—he combined his professional function with that of publican! In June that year he fell out with the resident agent and left the district. However, 'there is a very good surgeon not far from Coalcleugh, which many of the People have employed, and may be had on any occasion'.[21]

In the nineteenth century both the larger lead mining concerns established medical services, with compulsory membership for all miners. The Lead Company did so many years before the Beaumonts. The friendly societies provided some sort of income for sick men, but not usually free medical expenses. The London Lead Company's benefit fund, established in 1810, specifically allowed medical expenses. When it collapsed in 1827, and was replaced by a more directly company-controlled organisation, a separate 'health service' was provided. Surgeons were employed to devote all their time to looking after the company's workmen and their families without charge.[22]

The Beaumont concern provided a similar service some time after Sopwith became chief agent in 1845. Sick workmen and their families were attended by resident doctors in the different districts. It was not free; each workman paid 5s a year if he were married, 2s 6d if he were single. These charges covered some 45 per cent of the costs of the scheme, and the remainder was met by the concern. Unlike the Beaumont benefit societies, membership of the medical service was compulsory for all workmen.[23]

How much did the health of the miners improve? The complete absence of statistics for the eighteenth century makes the question impossible to answer. In the larger mines, at least, there were definite improvements in ventilation practice during the period. Improved ventilation at smelt mills undoubtedly lengthened the lives of the smelters. It is possible that the average age of death of the mid-eighteenth-century miner was nearer 35 than 45. However, the vital part played by mineral dust in causing pneumoconiosis was not recognised even by 1864, and there was no attempt to make workmen wear masks while drilling and blasting.

NOTES

[1] 1842 report (Mitchell), pp. 751–2.

[2] 1864 report (vol. 2), p. 362.

[3] Ibid (appendix), p. 434.

[4] Pages 63–4.

[5] Third report of the Medical Officer of the Privy Council, 1860, p. 144.

[6] 1842 report (Mitchell), p. 765.

[7] Ibid, p. 760.

[8] 1864 report (vol. 2), p. 337.

[9] Ibid (appendix), p. 12.

[10] Although certain levels and districts were recognised as being more fatal than others, sandstone was far more dangerous than limestone.

[11] W. Robinson, *Lead miners and their diseases*, 1893, p. 8.

[12] 1842 report (Mitchell), p. 755.

[13] 1864 report (vol. 2), p. 361.

[14] Ibid, p. 341.

[15] B/B 131, 23 June 1860.

[16] 1864 report (vol. 1), p. 281.

[17] 1842 report (Mitchell), p. 751.

[18] 1864 report (vol. 1), p. 261.

[19] Ibid (vol. 2), p. 359.

[20] Ibid (vol. 2), p. 353.

[21] B/B B.M. letter book, 5 January and 8 June 1733.

[22] See Raistrick, 1938, pp. 49–51, for a detailed account of the company's medical services, with long extracts from the rules governing the surgeons.

[23] 1864 report (vol. 2), p. 359; appendix, p. 19.

XI

Religion and social life

During the last ten years of the eighteenth century and the first three decades of the nineteenth, great changes took place in the religion and social life of the lead miners and their families. By this period the Church of England had become a very distant institution, in both the physical and the spiritual sense. There was a vacuum in the religious life of the area. Methodism filled this vacuum; the closeness and exuberance of the Methodist teachings, preached in their own homes by men of their own class in language they could understand, converted most lead miners to the new faith. This conversion and the puritan spirit it encouraged, combined with the mine owners' increasing economic efficiency and dominance over every part of their employees' lives, caused a corresponding reformation in the manners and social life of the area.

The Established Church

By the eighteenth century the Church of England, which had failed to keep pace with the growth of population, was not providing adequate services for its members in the large and isolated parishes of the northern Pennines. The places of worship were often too far away down the dales for easy attendance by the lead miners and their families, and burial of the dead in consecrated ground was difficult, the roads being so inadequate. In 1713 the incumbent of Hexham penned a graphic description of the difficulties faced by the people of the Pennine parishes:

The people here and in diverse other Places . . . go to the nighest Church or Chapel to Divine Service, some of them being compelled by a kind of necessity, to go . . . a begging the Bread of Life out of their own County and Diocese. . . . It is much to be wished, there were some effectual care taken, to remedy these Inconveniences, least they may prove a Temptation or Pretence to some, to go no where to worship God, others to despise Infant

Baptism, turn Quakers, and make themselves burying Places more convenient at home. For the carrying the dead upon Mens shoulders so many Miles, in such difficult way to the Parish Church, in some Places, in these Northern Counties, as Teasdale, Weardale, Hexhamshire, Allendale, etc., is such a heavy Burthen . . . as justly deserves to be redressed, and might effectually be done by enclosing diverse little Pieces of Ground, and having them consecrated for Christian Burial.[1]

The Blacketts were the patrons of the living of Allendale, and the incumbents were encouraged to pay special attention to the spiritual needs of the miners. A new perpetual curate was presented to the living in 1756, 'on condition of his . . . constant residence, and indulgence to the People of Allenheads and Coalcleugh'.[2] But the parish was so far-flung that the single parish church, at Allendale Town, was too far away for the miners living at the southern extremities of East and West Allendale. There were indeed two chapels a little further up, at Ninebanks, on the West Allen, and St Peter's in the Forest, on the East, but these were in a very decayed state by the early eighteenth century, and services were held in them only once a month. In any case, both places were still some five miles from the mines at Coalcleugh and Allenheads.

At the beginning of the century Sir William Blackett contributed the necessary timber for chapels to be built at Allenheads (1703, on the site of an earlier building) and Coalcleugh (1704). At each was established a chaplain technically independent of the parish structure who 'hath for his Salary half a Day's Wage of every Workman every month; which in time of Peace, when these Lead-Mines did flourish, amounted to between £70 and £80 a Year'.[3] This money was not always easy to collect; one Allenheads chaplain resigned in 1732 because he considered that 'he has been wronged of' about £70 in the previous six years by the agent's 'refusing to secure it for him from the People'.[4]

The new chapels existed rather uneasily within the old parish system, causing much jealousy and heartburning among the clergy. In 1731 some of the inhabitants of Ninebanks presented a petition to Sir Walter Blackett's agent that all miners in the area should pay 'half Shifts' to the curate who was officiating at Ninebanks chapel. The agent recommended to Sir Walter that the request be refused, as the petition was inspired by the curate, 'for whose ease and interest the thing is calculated'. Ninebanks was only two miles away from the parish church; 'all these Reductions are a burthen upon your works which ultimately must be paid by you. Allenheads, Coalcleugh are expected to pay shifts for very good Reasons, not one of which will be held in the present case.'[5]

In 1764 the situation was further complicated by the rebuilding of Ninebanks chapel and the granting of a licence for a burial ground there. A permanent curate was appointed who was independent of the parish church, the incumbent of which had no say in his appointment. In the 1790's there was a bitter quarrel between the chaplain at Coalcleugh, Joseph Carr, who was also the incumbent of Allendale parish itself, and the curate of Ninebanks, Nicholas Richardson. In the Hexham manorial papers there is a lengthy 'impartial account' (undated) of this quarrel, submitted as a petition to Mrs Beaumont by Joseph Carr. The petition well illustrates the confusion that existed within the established Church in Allendale.

Joseph Carr stated that until the re-endowment of Ninebanks in 1764 the Coalcleugh chaplain 'by ancient custom' performed certain duties at Coalcleugh for the miners in addition to holding services. These included the churching of women and the baptising of children, for neither of which was any charge made. Marriage and burial rites had to be carried out at the parish church, seven miles away. It was largely to lessen the distance for the miners on these occasions that Ninebanks chapel was re-endowed, the curate there being able to both marry and bury the inhabitants of Coalcleugh. But Mr Richardson, ever since his appointment in 1772, had been trying to make the miners 'pay him for duty done gratis by their Chaplain . . . In case of refusal he threatened them with prosecutions in the Spiritual Court'. Mr Carr had told the miners not to pay any dues, and Mr Richardson had complained to Mrs Beaumont. Carr concluded his rebuttal of Richardson's charges by writing as perpetual curate of Allendale rather than as chaplain of Coalcleugh.

He begs leave further to add, that the separation of this Chapel from the Parish Church has been, and still continues a disadvantage to the Incumbent of the Parish, as well as a loss to the Patron. . . . In all cases, in general, where the Parish Church is poor, as at Allendaletown, the Incumbent is made Patron of the Chapels in his Parish and by these means, the Patronage of the Parish Church is virtually increased, but in this case, neither can the Patron of the Parish Church get any benefit, or the Incumbent augment his income thereby, since neither has any power over it.

Carr had some justification in complaining about his poor income for, according to the Allendale mining agent, writing to Colonel

Beaumont in 1795, 'after paying his curate at Allendale I don't believe he has above £40 per Ann.; out of which he is under the necessity of keeping a Horse as he does Duty at a distant Chapel.'[6] If Allendale was typical, the Church of England was not in a flourishing state in the northern Pennine dales. The church accommodation was inadequate and inconvenient and the clergy were poorly paid—although the parish of Stanhope was an exception here.

When the mines were flourishing the rector of Stanhope, by virtue of being, with the Bishop of Durham, joint owner of the mineral rights, was probably the best-paid parish priest in Britain. In 1835, the year before the Commission on Ecclesiastical Revenues reported, the rector received a gross income of £4,875. But the benefice was customarily held in pluracy by a non-resident priest, often a bishop. In 1835 the two curates resident in the parish were together paid a total of £279 out of the rector's vast income. Little if any of the income was spent within the parish.

The Church was in as poor a state in Weardale at the beginning of the nineteenth century as it was in Allendale. An undated petition of about 1800 lists the ecclesiastical grievances of some of the inhabitants of St John's Chapel, in which chapelry 'nine-tenths of the whole Inhabitants of the Parish' lived:

That in Consequence of these Circumstance your Petitioners have the strongest Reasons to believe (but which Delicacy will not permit them to state) that, altho' the Population of the Parish is not in any Degree diminished, yet that Marriages are not so frequently celebrated as would be the Case, was the Right of Marriage annexed to the said Chapelry of St. John. . . . The burying ground belonging to the said Chapel . . . is by far too small for the internment of the Inhabitants and . . . on every Funeral the most indecent Violence is unavoidably done to the Remains of the Dead already deposited there.[7]

Nonconformity

There is little evidence about Nonconformity in the region before the arrival of Methodism. In the late seventeenth century there was a small and persecuted group of Quakers in Allendale. Persecution lessened in the eighteenth century, but by 1800 the Society of Friends was virtually non-existent in the dale.[8] Presbyterianism survived on Alston Moor and in Weardale until the end of the eighteenth century. At the beginning of the century Adam Wilson, of the Presbyterian Church of Scotland, became minister at the Independent chapel in

Garrigill.[9] He founded another chapel at Ireshopeburn in Weardale before 1720.[10] In 1769 Sir Walter Blackett contributed £10 towards a new meeting house in Weardale. 'It is represented that this meeting house is in the midst of many of Sir Wr's Miners, that are of the Church of Scotland, and therefore Sir Walter's willing to accomodate them with a place for orderly worship, rather than that Religion should be neglected among them.'[11] By 1840 this chapel had been turned into a school, and no members of the Church of Scotland were left in Weardale.[12] Roman Catholicism was effectively non-existent at all times in the dales in the eighteenth and nineteenth centuries.

Methodism reached the lead mining area in the late 1740's. Wesley himself first preached in the area in Blanchland, on 24 March 1747, recording that his large congregation 'gathered out of the lead-mines from all parts; many from Allendale six miles off'.[13] The practice of the early Methodist preachers was to travel over the country to preach wherever large gatherings of people might be expected; Wesley's visits seem to have often coincided with the pays of the mining companies. Having sown the seed, the leading Methodists would travel on, to reappear some months or years later to revitalise their following. Wesley himself, of course, travelled the whole of England and Scotland in this way. Lesser men travelled more localised circuits. Enthusiasts from one dale would cross over to the next to spread the Word. One early northern preacher (Christopher Hopper, the first to enter Allendale, in July 1747) described his approach to a new area:

Our plan was to visit a town or a village, and ask permission to expound the word of God in one of their houses or cottages; if the people did not invite us to lodge and break bread with them, after repeating our visit two or three times, we took it as an indication, that we were not called to such a place.[14]

It does not appear that there was any very violent reaction to Wesleyan missionary efforts in the area from either the civil or the religious authorities. The only recorded persecution was in Teesdale, where, according to Wesley in 1761, many miners 'a while ago were turned out of their work for following "this way". By this means many of them got into far better work, and, some time after; their old master was glad to employ them again.'[15] The reception accorded to Wesley himself was normally good. In Blanchland in 1748 'at the desire of Mr. W., the steward of the lead mines' he preached in the house where the pays were taking place.[16] In Allendale Town, two days later, he disputed publicly with the Church of England curate, but the curate 'skipped so from one point to another, that it was not possible to keep up with him' and so the disputation was abandoned.[17]

In these early days Wesley and his followers were aiming at a revitalisation of the Church of England, not at the formation of a new Church. Hence no organisation was created to follow up the work of the itinerant preachers. The converts met in each others' homes for prayer meetings, but frequently their enthusiasm disappeared with the preacher. Thus, after his highly successful series of meetings in Allendale in 1748, Wesley found, when he returned in 1752, 'the poor society well nigh shattered in pieces. Slackness and offence had eaten them up.'[18]

The Methodist campaigns in Weardale are particularly well described in his journal. He went there more frequently than to the other dales, having a particular liking for the scenery and the people, and in two long entries he gives a brief history of the rise of Methodism in Weardale, and an analysis of the reasons for a great decline in membership in 1773–74.[19]

Weardale was originally converted from Allendale. Two preachers went from Allendale to Weardale in 1749, and as a result of their efforts four inhabitants 'found peace with God and agreed to meet together'. By 1750 the society had about twenty members. It then gradually increased to about thirty-five, and remained around that number for ten years. Wesley noted that 'they have been particularly careful with regard to marriage', marrying only with each other. In 1761 he visited the dale 'just in time to prevent their all turning Dissenters, which they were on the point of doing, being quite disgusted at the curate, whose life was no better than his doctrine'.[15] In November and December 1771 there was a great increase in membership. Meetings were held all over the dale, and there were many public and emotional conversions. Of the new members, who were some 120 in number, over forty were children, their schoolmistress being a Methodist.

Two and a half years later, however, Wesley found a 'grievous decay in the vast work of God'. He asked to see one girl he had himself converted; she 'was brought almost by force. But I could not get one look, and hardly a word, from her.' In his journal he noted what he considered the five reasons for the movement's sudden collapse in Weardale: (1) the local preachers were poor and uninspiring; (2) the school-teacher had married, and her ex-pupils, 'there being none left who so naturally cared for them ... fell heaps upon heaps'; (3) 'most of the liveliest in the society were single men and women, and several of these in a little time contracted an inordinate affection for each other, whereby they so grieved the Holy Spirit of God that He in great measure departed from them'; (4) there was doubt whether such rapid conversions as had taken place could be really sincere; (5) there had been much discontent and bickering between members.

In 1757 the Dales circuit was founded. Based on Barnard Castle, it ensured a regular circulation of preachers throughout the northern Pennines. Some statistics of membership were kept, which show that although there was a rise in membership between 1760 and 1770— from fifty-seven to eighty in Teesdale; from thirty-one to thirty-eight in Alston—the only place with a dramatic increase in numbers was Weardale, and we have already seen what subsequently happened there.[20] Fluctuation in numbers was normal even later in the eighteenth century. Some detailed figures for individual areas in upper Teesdale survive for the years 1791–1800. Middleton itself had twenty members in 1791 : twenty-eight in 1794; and only ten in 1800. They show fairly conclusively that it was not until the nineteenth century that Methodism attained the dominant position in the life of the dales that it possessed in 1851 when the religious census showed that in every dale there were more Methodists than members of the Church of England. Of Weardale, where Methodism was the 'established religion of the dale' in 1840,[21] it was said by Eden in 1797 that 'the inhabitants are chiefly of the Church of England; but there is one congregation of the Methodists, and one of Presbyterians'.[22]

By the mid-nineteenth century the predominance of the Church of England had completely disappeared. The numbers of Wesleyan Methodists increased steadily, and from 1820 onwards they were joined by another proselytising sect, the Primitive Methodists. In 1840 the minutes of the Committee of Council on Education gave statistics of churches and church attendance in Alston parish. These show that of a total population of 6,062 (1841 census figures) there were *average Sunday congregations* of 1,030 Wesleyan Methodists, 700 Primitive Methodists, 250 Independents and ten Quakers. The figures for the Church of England were not given. In the 1851 religious census, the numbers attending evening service on *one particular Sunday* in Alston poor law union were 357 in the Church of England, 663 Wesleyan Methodists, 892 Primitive Methodists, and 147 Independents. Thus in these eleven years the Primitive Methodists seem to have gained at the expense of the Wesleyans and Independents. In Teesdale in 1851 there were similarly more Primitive than Wesleyan Methodists. Only in Weardale, the only dale where attendance at the Church of England had sunk below 100, were there more Wesleyans than Primitives.

In the nineteenth century the spread of Methodism in the mining area is less well documented than it is in the eighteenth, the period on which the Methodist historians concentrated. But it was in the first half of the century that the great expansion occurred. In 1797, the Church of England had been dominant in Weardale; in 1851 there were 1,441 people attending Methodist chapels, compared with seventy attending the Established Church. Featherston, writing in Weardale in 1840, summed up the changes that occurred in his lifetime, due mainly, he said, to the influence of Wesleyan Methodism:

They have been . . . the principal engine in effecting a moral change in this wild district, and instead of insult and a volley of stones, strangers are met with civility and good behaviour. . . . They have reclaimed and reformed individuals who were enemies to their families and themselves, as well as a perfect pest and a disgrace to the neighbourhood.[23]

The success of the Primitive Methodists was even more rapid. Their 'love-feasts' possessed greater appeal to many people than the more sober services of the Wesleyans. Featherston, a Wesleyan, remarked rather drily, 'it is a serious question for these serious people, if night meetings, particularly in the winter season, be for good or be for ill to the morals of young men and women, if unaccompanied by their parents and friends.'[24] The movement spread into the area in the early 1820's, resembling Wesleyan Methodism in its mode of propagation; travelling preachers provided the initial impetus, followed by local efforts to maintain membership. In Weardale the society was founded in November 1821; in March 1823 there were 219 members and by December 1823 there were 846. Enthusiasm and membership fluctuated wildly, but the 1851 religious census shows that in Alston Moor and Teesdale the Primitive Methodists were probably more numerous than the Wesleyans. One of their preachers described an open-air meeting in Weardale in 1824:

While many spoke of the goodness of God, a mighty power came down. It struck one (a believer) speechless; two others fell to the floor in great agonies, and rose praising God for what they felt. Another man began to pray for a clean heart . . . and soon after he was so filled with the perfect love of God that he jumped up and down, shouting 'Glory' with all his might. . . . Sinners then began to tremble before God, and presently five or six fell down and cried for mercy.[25]

The Church of England could offer little to compare with such enthusiasm. Its vicars and curates were of the upper and middle classes, without the sympathy for their parishioners the Methodist preachers, who were often ex-miners themselves, possessed. Sopwith wrote of the attendance in Allenheads chapel at a Sunday morning service in 1868. It consisted of 'from twenty to thirty persons chiefly members of families connected by Agencies or other occupations with Mr. Beaumont's works'.[26] The population of Allendale itself, he wrote on another occasion, was 'nearly equally divided between Wesleyan and Primitive Methodists. In few places is the influence of Dissent greater —or that of the Church of England less.'[27] In Weardale in 1840, 'Churchmen are so thinly sown' that 'many a time the Wardens are either Methodists or Ranters [Primitive Methodists]'.[28] Church accommodation in the nineteenth century was very inadequate compared with the chapels of the Methodists. In Alston parish, the population of which was 6,062 in 1841, there were two churches in 1840, with a total of 950 seats; six Wesleyan Methodist chapels with 1,910 seats; and four Primitive Methodist chapels with 1,155 seats.

Of other sects in the lead dales little is known. In 1851 the Independents were more numerous in Alston than adherents of the Church of England. The Society of Friends was unrepresented, save by a small congregation in Alston. It seems that the London Lead Company directors made no attempt to proselytise their employees. In 1840 there were a few Mormons in Alston parish, converted from Canada in 1837, but the sect apparently did not thrive.[29]

With a single exception there is no evidence that the growth of Methodism increased social friction between the employers and workers, despite the polarisation of classes around church or chapel by the mid-nineteenth century. The exception was the bitter Allenheads strike of 1849, where the leaders were Primitive Methodists and the chapel was used as their rallying place. Nine years after this event, however, in 1858, W. B. Beaumont on Sopwith's initiative began to make an annual donation to both the Wesleyan and the Primitive Methodists in Allendale and Weardale.[30] Even before the 1849 strike Sopwith had ended, against the strong opposition of the Allenheads chaplain, the enforced attendance of all children of whatsoever denomination at the Church of England Sunday school.[31] The London Lead Company impartially enforced the children at its schools to go to church *or* chapel on Sunday: 'Steady persons are stationed by the Company's agents at the door of every place of public worship, from whom the pupils of the Sunday schools receive tickets on entrance.'[32] Methodism was not seen as a revolutionary threat by the mining employers.

Poaching

Serious crime was never common. Dismissal and consequent loss of livelihood were a greater deterrent than legal penalties. With the exception of poaching, the subject is hardly ever mentioned in the correspondence of the Blackett/Beaumont concern or of the Greenwich Hospital in the eighteenth century. It is true there were sporadic outbreaks of lead stealing, but when this did occur it was usually embezzlement on the part of a carrier, rather than outright theft, in spite of the pigs of lead being left 'lying by the road sides and in the fells as much exposed as so many stones'.[33] In 1842 it was said that the office of constable in the parish of Alston was virtually a sinecure: 'There have been no riots for very many years, and very few breaches of the law of any sort; robbery and housebreaking are now never heard of.'[34]

The one great exception to this law-abiding record was poaching, which in a moorland district was bound to be a serious temptation both as a sport and as a supplementary food supply. The eighteenth-century letters from the Blacketts' chief agent in Newcastle frequently contained admonitions to the local agents to warn their men of the consequences of being caught at it, but poaching continued, for in bad times grouse and snipe were an invaluable source of food for the miners:

Now the times being hard and provisions being dear,
The miners were starving almost we do hear;
They had nought to depend on, so well you may ken,
But to make what they could of the bonny moor hen.

There's the fat men of Oakland, and Durham the same,
Lay claim to the moors, likewise to the game;
They sent word to the miners they would have them to ken,
They would stop them from shooting the bonny moor hen.

O these words they were carried to Weardale with speed,
Which made the poor miners to hang down their heads;
But sent them an answer they would have them to ken
They would fight till they died for their bonny moor hen.

The verses are from a famous ballad recounting an exploit that took place in the lean year of 1818 when the Bishop of Durham's men arrested a number of Weardale miners who were subsequently forcibly released by their fellows.

The early 1820's were bad years for landlords near the lead mining area. The steward of the Wallace estates at Featherstone castle, to the north of Alston, had much trouble from frequent poaching by large bands of miners. He had difficulty in supplementing his staff of game-keepers, as local men 'would be afraid to go to Alston afterwards'. In one incident

upwards of 100 people were assembl'd to take 12 poachers. and they bid defiance to them all, they plac'd themselves back to back, threw down their Game, and dar'd any of them to come within a certain distance, if they did, 12 of them should be dead men that instant, and consequently they were suffer'd to go away without any further molestation.

He thought of writing to the London Lead Company's agent but 'I am told applying to the Company will be of no use and that some of the Poachers are employ'd to shoot Game for them.' A neighbouring landlord

sent some summon's the other day by a Constable to serve upon some Poachers in Garrigill who had been seen upon his Manor . . . but he was oblig'd to return without serving them, about 40 of the Poachers assembled as soon as it was known he was there, and he was glad to make his escape.[35]

After these stirring events the poaching lessened, the miners being alarmed by the threat of prosecution.

Such outbreaks of violence were apparently the last large-scale acts of poaching by large bodies of men. Later in the century poaching became more the purview of individuals. Most miners were dis-couraged by their new Methodist faith, and by the increasing efficiency of the mining companies in searching out misdemeanours and punish-ing them by dismissal.

Drinking habits

The influence of these two factors—the growth of Methodism and the increasing paternalism of the companies—is also seen, and that most strikingly, in their effect on the drinking habits of the miners. In 1800 the Alston men were said to 'work hard about four days in the week, and drink and play the other three'.[36] In the eighteenth century strong drink was the favourite consolation of most miners. The letter books of the Blackett concern show that the provision of alcohol was condoned by the company, although care was taken that there should not be too many places where it could be obtained at the mines themselves. In the 1750's the Coalcleugh agent received an allowance for supplying his men with ale at the letting of bargains.[37]

Earlier, in the 1720's, drink was regularly supplied to men working in one of the Derwent mines when specific jobs were completed. In 1733 the man who kept the public house at Allenheads left the district, and there was some discussion between the chief agent and Walter Blackett as to who should be permitted to succeed him in this lucrative position. 'There are choice of Candidates for the House, but to let it any more to a Trader, experience shows, is neither for yours nor the Workmen's interest. I mean not to lett it to a shopkeeper, for I think its very necessary to have a publick house there, if it can but be kept under proper Restriction.' It was not advisable, however, to let it to the son in law of the Allenheads agent, who was one of the applicants.[38] In 1795 the then chief agent advised Colonel Beaumont to close an alehouse at Allenheads formerly kept by the recently dismissed second agent, Westgarth Forster. The alehouse had been opened in spite of the existence of the

old accustomed public house kept by Nicholson the Engine Man. . . . It has been a rule laid down by the late Sir Walter Blackett . . . not to allow more than one public house at Allenheads for reasons too obvious to mention. The breaking thro' this has been attended with very bad Consequences not only to the Concern but to the Health and Morals of the Workmen.[39]

The men drank on every social occasion connected with their work. After the letting of bargains and the payment of subsistence the miners would 'adjourn from the office to the public-house, and it not infrequently happens that a battle ensues before they part'.[40] The washer boys were introduced early to the custom. Sopwith recounts a pleasant little story about the drinking habits of the Allendale washer boys, worth quoting because of the light it throws both on these customs and on the Victorian reaction to them:

In two successive years I found that on a certain day about midsummer many of the washing boys . . . were intoxicated—some loudly quarreling—others lying about apparently insensible. On enquiry I found that on the 29th June it was usual for these boys to be indulged not only with a holiday (i.e. Holyday!!!) but also with superabundant refreshment leading to the above results.

This, and this only, was the last remnant (for I destroyed it) of the annual feast or festival of St. Peter the tutular saint of the Parish.[41]

By the 1840's there had sprung up many forces opposed to excessive drinking, or indeed to any drinking at all. The Methodists of course regarded alcohol as the root of most evils. A medical witness remarked in 1842 on the extensive and effective teetotal propaganda being circulated among the miners, and many witnesses testified that the amount of alcohol consumed in the area was much less than in former times. This beneficial development, commented the commissioner, was 'caused partly by poverty as well as principle'. More than half the Weardale miners he interviewed were teetotallers. The inns kept soft drinks for the abstainers who came to enjoy 'the sociality of the public-house'—'coffee, soda-water, ginger beer, lemonade, black-beer and peppermint'. The concentrated solution of peppermint was the most popular, and cost the same as rum; one landlord regretted 'that poor men should be deluded to spend their money on such stuff, when they might get good wholesome invigorating beer'.[42] But the inns themselves were disappearing. Allendale had seventeen public houses in 1827.[43] By 1863 there were only two in the whole of East Allendale, while in West Allendale 'we have had the people petitioning, for the only public house at the principle village of the district to be done away with'.[44]

At least as important a cause as the growth of the temperance movement was the attitude of the mining companies. In the eighteenth century the Blackett/Beaumonts did not attempt to do more than regulate the consumption of drink. There is no evidence of the London Lead Company's attitude but it was probably similar. Eden states that in 1797 there were no fewer than forty-six alehouses on Alston Moor.[45] By 1832, when a surplus of labour enabled the company to make its employees obey almost any regulation it chose to lay down, 'strict discipline' had been established 'for the prevention of drunkenness and debauchery'. A first offence was visited with an admonition or a fine, the second by dismissal.[32] By 1842 Lead Company employees were prohibited from entering public houses, and the penalty for drunkenness was instant dismissal.[46] The Beaumont concern had followed the company's example by 1861, when Sopwith noted in his diary the dismissal of six men for drunken rowdiness in a public house, although regretting that it meant they had to leave the district.[47] The London Lead Company, incidentally, declared war upon other vices

too. In 1842 swearing was punished by fines of 1s for a man and 6d
for a boy, to be paid into the benefit fund.[48] An employee found guilty
of fathering illegitimate children was made to marry the mother or be
dismissed.[32]

Sports and pastimes

There was a similar toning down of sports and pastimes. In the eighteenth century the rougher blood sports were common—cock and dog
fighting, badger baiting—as well as the milder ones such as wrestling
and hare hunting. The early Methodists preached at cock-fights—the
places where they found the greatest congregations of people. By the
1840's cock and dog fighting had disappeared, and even wrestling was
dying out, though hare hunting and hound trails continued. 'A hound
is as requisite to complete a miner's establishment at one stage of his
life as a wife,' wrote Featherston in 1840.[49] Sopwith's diary records
his horror at discovering that his position as Allenheads agent required
him to be 'master of the hounds' and lead his men in pursuit of the
hare.[50] He soon contrived to pass on this honourable duty to the chief
smelting agent, who performed it until after his eightieth year. Before
Sopwith left Allendale in 1871, hare hunting too had almost vanished
from the scene. Horse races were held regularly at such centres as
Alston, Stanhope and Allendale Town in the eighteenth and early
nineteenth centuries, but by 1850 the races had been abandoned at
all three.

The more disciplined amusements that replaced the blood sports of
the eighteenth century included cricket, which both the Beaumonts
and the Lead Company supported generously, and a volunteer corps
in each of the dales, again well patronised by the mine owners. Each
tiny community also had its own town band. As well as being a
symptom of the restraining and civilising influences of Methodism,
the changes illustrate the greater power the mining employers had
gained over all aspects of their employees' lives.

Another feature of social life which was declining by the mid-
nineteenth century was the great fairs and merry-makings which had
formerly accompanied the pays and annual hiring fairs. Agricultural
and horticultural shows took their place. Even dancing had declined,
and was scorned as a worldly amusement by the Methodist preachers.
Also disappearing was the custom of holding great feasts to celebrate
funerals as well as weddings.

Thus the social life of the lead miners changed with the gradual triumph of Methodism and the increasing economic power of the mine owners. In the coal mining areas, too, Methodism acquired a dominant position, but there the economic power of the landlord was not so great. The coal miners organised themselves into unions to fight their employers, and if they were discharged there were nearly always other mines ready to take them on. Rough sports and riotous conduct continued well beyond the mid-nineteenth century. A well known *Punch* cartoon of 1854 entitled 'Further illustration of the mining districts' shows two coal miners watching a well dressed man approaching them. Says one, 'Who's 'im, Bill?' 'A stranger.' ''Eave 'arf a brick at 'im.' It is symbolic of the mid-Victorian middle-class attitude to the coal miner—a savage outside the pale of civilisation. The Victorian commissioners and others who visited the lead mining district found the state of affairs there very different. In 1859 a guide-book writer commented, 'The miners are, for the most part, sober and industrious; there appears to be something in their metalliferous employment which makes them, as a class, more respectable than coal-miners.'[51] It was not the metalliferous nature of their employment, nor even the influence of Methodism, which produced this appearance of sober respectability but rather the grimly paternal discipline insisted upon by their powerful employers.

NOTES

[1] Ritchel, pp. 66–7.

[2] B/B 47, 8 April 1756.

[3] Ritchel, pp. 16–17.

[4] B/B B.M. letter book.

[5] B/B B.M. letter book, 21 November, 3 December and 8 December 1732.

[6] B/B 50, 28 January 1795.

[7] Thomas Bell MSS.

[8] Dickinson, pp. 89–100.

[9] J. H. Colligan, *Nonconformity in Cumberland and Westmorland*. Church Historical Society publications, 1907, p. 214. The Scottish Presbyterians took over many Independent congregations in the north of England after 1688. This does not imply massive Scottish immigration.

[10] T. S. James, *The history of . . . Presbyterian chapels and charities in England*, 1867, pp. 654–5.

[11] B/B 48, 28 December 1769.

[12] Featherston, pp. 18–19.

[13] *Journal*, 27 March 1747.

[14] Quoted in A. Steel, *History of Methodism in Barnard Castle*, 1857, p. 105.

[15] *Journal*, 9 June 1761.

[16] Ibid, 27 July 1748.

[17] Ibid, 29 July 1748.

[18] Ibid, 26 May 1752.

[19] Ibid, 4 June 1772 and 12 June 1774.

[20] Steele, pp. 118, 189.

[21] Featherston, p. 30.

[22] Eden, vol. 2, p. 168.

[23] Featherston, pp. 31–2.

[24] Ibid, p. 47.

[25] W. M. Patterson, *Northern Primitive Methodism*, 1909, pp. 161–2.

[26] Sopwith, *Diary*, 5 July 1868.

[27] Ibid, 4 October 1863.

[28] Featherston, p. 70.

[29] Minutes of the Committee of the Council on Education, 1840–41, p. 149.

[30] Sopwith, *Diary*, 19 November 1858.

[31] Ibid, 9 March 1847.

[32] Poor Law Commission report, 1834, appendix, A p. 139A.

[33] 1842 report (Mitchell), p. 752.

[34] Ibid, p. 767.

[35] Wallace letters, Northumberland Record Office, 7–25 September 1820.

[36] Housman, p. 70.

[37] B/B 98, 30 September 1753.

[38] B/B B.M. letter book, 5 January and 8 June 1733.

[39] B/B 50, 24 September 1795.

[40] Featherston, p. 64.

[41] Sopwith, *Diary*, 25 August 1856.

[42] 1842 report (Mitchell), p. 753.

[43] Parson and White.

[44] 1864 report (vol. 1), p. 264.

[45] Eden, vol. 2, p. 89.

[46] 1842 report (Mitchell), p. 758.

[47] Sopwith, *Diary*, 20 August 1861.

[48] 1842 report (Mitchell), p. 765.

[49] Featherston, p. 15.

[50] Sopwith, *Diary*, 27 September 1845.

[51] White, p. 49.

XII

Education

'As to the intellectual condition of the people, it is decidedly superior
to that of any district of England of which I have any knowledge.'[1]
This comment was one of several like it made by Assistant Commissioner Mitchell in 1842. A. F. Foster, the assistant commissioner who
reported on the mining area for the Newcastle Commission on Popular
Education of 1861, also found that 'the lead miners are remarkably
intelligent, and generally well educated'.[2] Other nineteenth-century
writers expressed themselves in a similar manner, contrasting the lead
miners favourably not only with the neighbouring agricultural and
coal mining communities but also with 'labouring classes' anywhere
else in England.

This commendable state of affairs was due to a number of inter-
related factors. Relatively short hours of work and the cessation of
washing activities in winter allowed time for education. Methodism
was an important civilising influence. The mining environment too
seems in some nebulous way to have encouraged intellectual pursuits.
Lastly, the region's schools and libraries, little better or worse than
those of other rural areas in the eighteenth century, improved in
number and quality after the Napoleonic wars. At the very end of the
eighteenth century Eden commented that the lead miners were 'less
profligate' than coal miners, and 'mostly better informed'.[3] At this
time the schools in the region were certainly in no way superior to
those in the coal mining areas. In 1821 a Greenwich Hospital report
said much the same of the Hospital's mining tenants on Alston Moor,
explaining that 'the nature of their Occupations as Miners, leads them
to enquiries, which greatly quicken their understandings and urges
them to seek from Books such facts of practical Philosophy as are
applicable to their profession.'[4]

All miners needed to understand the geology of the area if they were
to be successful, and the actual work of extracting the ore using
explosives required skill and courage. Hydraulic and other machinery

was assembled and often designed by the miners. Most important of all, however, in stimulating intelligence was the complex bargain system governing employment. A miner needed to know how to estimate the potentialities of the ground he was working, and to be skilled at calculating and bargaining. For the latter an elementary knowledge of arithmetic and accounting was a great asset. The miners 'generally possess a degree of shrewdness and intelligence rarely found in a labouring class of people,' wrote Sopwith in 1833.[5] The smelters and refiners also required a high degree of mechanical and chemical skill to carry out their work.

The advantages of possessing an elementary education as a means of acquiring these skills were as obvious to the eighteenth-century miners as to their nineteenth-century successors. Even in the earliest bargain books of the Blackett/Beaumont concern, dating from 1720's, a bargain signed 'X John Doe his mark' is uncommon. Nearly all were signed with a proper signature, and the same man did not necessarily sign for the whole partnership every time a bargain was renewed. In 1842 literacy was not a new phenonomen; 'children in the lead country have the benefit of the example of their parents, and their encouragement to attend to their education; whilst in many other districts the fathers, never having learnt to read themselves, make no effort to stimulate their sons'.[6] In 1861, books were to be found in nearly every home. Some miners wrote books about their profession. Westgarth Forster, the author of *A section of the strata from Newcastle upon Tyne to Cross Fell* (first edition, 1809), a work which became a geological classic, was a mining agent on Alston Moor, the son of another at Allenheads. John Leithart, who wrote *Practical observations on . . . mineral veins: with the application of several new theoretical principals to the art of mining*, published in 1838, stated in his introduction that 'a large portion of his time, from his infancy, was spent in the mines. He had not the good fortune to obtain the usual advantages of education . . . A Sunday School . . . was the only school . . . that the author ever attended.'

The educational history of the area up to the 1870 education Act divides up fairly logically into three periods of unequal length. In the eighteenth century and in the nineteenth up to about 1818, schools led a precarious existence, supported inadequately by private charity with a little help from the mining companies. In the second period, from 1818 until the early 1840's, occurred the foundation of the London Lead Company schools, originated by the reforming zeal of the company's Quaker chief agent, Robert Stagg. At the same time there was a great increase of support for other schools in the area and the foundation of new ones by the Church of England, by the ground landlords and by other mining companies. Lastly, during the late 1840's and 1850's educational facilities were greatly improved in the Beaumont mining districts, owing to the work of Thomas Sopwith. A. F. Foster's report to the Newcastle Commission on Popular Education, published in 1861, contains lengthy accounts of the schools of the two chief mining companies, with comparisons between them.

Schools

Educational institutions in the eighteenth century existed under the same conditions and suffered the same disadvantages as the Established Church, with which nearly all were connected. The parish areas were enormous and the inhabitants' dwellings scattered. A single centre in a parish was not accessible from all parts of it. The eighteenth-century schools were charity schools, endowed with land and/or a regular income. This income was rarely adequate, and either the schools had a minimum number of pupils, if they existed *de facto* at all, or fees were charged to supplement the teacher's salary. There were probably also dames' schools, the teachers being little more than child minders.

In each of the chief mining parishes—Allendale, Alston, Blanchland, Middleton and Stanhope—there was at least one charity school in existence in the eighteenth century. A typical institution was that in Garrigill, founded by a legacy left in 1685, which specified 'Twenty Shillings a Year to a Schoolmaster at Garrygill, towards his Maintenance; and forty shillings a Year to the said Schoolmaster for teaching six poor Children of the poorest Inhabitants in Garrygill, gratis, till they can read the Bible through, and then others to be put in their stead'.[7] Alston town had a grammar school. In Middleton parish there were three charity schools founded in the eighteenth century, one in Middleton itself (from 1729) and two in outlying settlements to the east of the parish: Harwood (1724) where the incumbent of the Church of England chapel of ease received £4 a year from an endowment for teaching a number of poor children free, together with another £5 from the Lord Crewe charity; and Newbiggin (1799) where a teacher received £11 annually from an endowment to pay his salary and for fuel. The first and last of these three schools accepted paying pupils in addition to those taught free under the terms of the charity. In Stanhope parish there were four charity schools, three of them in the centre or eastern parts of the dale, whereas most lead miners lived in the west —Stanhope (1724), Westgate (1681), Boltsburn in Rookhope (1762) and St John's Chapel. In Blanchland a school was founded by the Lord Crewe Charity about 1750. In Allendale Town a school was endowed with both land and income in 1692 and 1700. Early in the nineteenth century this income amounted to about £50; the master was not allowed to take any fee-paying pupils, and he sometimes had to teach as many as 250 children single-handed. At Ninebanks on the West Allen the incumbent of the chapel received £1 a year 'for his encouragement and provision to teach a school at the said chapel'.[8]

Endowments existed for these charity schools. Whether the schools themselves date from the time of the endowment is less certain. At Blanchland, for example, the difficulty of finding a suitable teacher— or even a teacher at all who would work for the money available—was never really solved in the eighteenth century. The post was vacant intermittently for considerable periods over a number of years. In a petition from the inhabitants in 1778 one teacher was alleged to be incapable, children 'who went many years to that school, not being able to read or write fit for business'. This particular teacher was in prison for a time for theft.[9] Situations like this probably occurred in many schools in the area.

In 1805 three of the directors of the Greenwich Hospital included in their printed report of an inspection of the Hospital's estates in the north a survey of schools on Alston Moor giving more details than the customary lists of endowments.[10] There were five schools, altogether. The largest was Alston school, founded as a grammar school in the sixteenth century. In 1805 it had seventy-two boys, 'three of whom have a classical education; twenty-eight are taught reading, writing and arithmetic; and thirty-one reading only.' The vicar was the head of the school, with an assistant master. The school's endowed income was £16 7s with £53 4s given by the boys' parents in the form of quarterly payments. From this the vicar received £39 3s annually, and his assistant, who probably did most of the work, £30 8s.

Nenthead school was erected 'about 30 years' before 1805, by voluntary subscription. There were then twenty-seven boys, fifteen being taught reading, writing and arithmetic, and twelve reading only. The master received £40 a year, 'which was intended to be raised by small monthly payments from the parents ... of the boys, but they always fall short; this School is patronised by the Governor and

Company and the deficiences in the monthly payments are made good by them or their agents.'

The charity school at Garrigill had forty-four boys. These included the six poor children of the poorest inhabitants' being taught to read free. Thirty-two more were taught reading only, and six writing and arithmetic also. The master's income was £27 6s a year, £7 8s from endowments and the remainder from monthly payments by the parents.

Nenthall school, built in 1789 by voluntary subscription, contained thirty-five boys divided into 'twenty-six readers, six writers, and three arithmetical scholars'. The income was £20 4s a year, £18 14s from quarterly payments by the parents, and £1 10s from the rent of a cottage over the schoolroom. The last school, at Leadgate, was 'erected by voluntary contributions about thirty years ago'. Of the thirty-one boys, fourteen were taught reading, twelve writing, and five arithmetic.

The London Lead Company supported the Nenthead school, and the master's income there was higher than at any of the others. In giving this support the company was following the practice of the Blacketts at their mines at Allenheads and Coalcleugh. In 1703 Sir William Blackett started making annual payments to a schoolmaster at each of these places—£10 a year at Allenheads and £5 a year at Coalcleugh—in addition to the quarterly or monthly payments by parents of pupils.[11] The schools continued into the nineteenth century. At Coalcleugh, sums paid to the master every quarter were recorded in the official quarterly accounts of the mine. At Allenheads it became customary for the chaplain to be responsible for teaching in the school, and the annual subscription of the produce of four days' work per year from each workman due to him for his spiritual duties became conditional on his teaching in the school. As late as 1840, when the Allenheads chaplain was also the incumbent of Allendale parish, he was still nominally responsible for the school, and inspected it daily.[12]

At Langley mill in 1780 the Greenwich Hospital ordered a school to be built for the new settlement there. In 1783, however, the building so constructed was converted into a cottage.[13] Unfortunately the letter book recording this latter decision has become illegible through damp, and the reason for it cannot be deciphered.

This survey has shown that in the eighteenth century the schools were provided chiefly by the private generosity of individuals, together with efforts by parents to give their children some sort of an education.

The mining companies assisted these efforts to a very limited degree. Their support of schools was probably due less to educational zeal than to a desire to keep the children out of mischief and inculcate some sense of discipline. In 1796 the Beaumont chief agent wrote to Colonel Beaumont about the school at Allenheads:

All that is necessary for the Boys to be taught is reading, writing, and the common rules of arithmetic. . . . When the boys are instructed in Mensuration, Trigonometry, etc., it frequently unfits them for Miners; they sometimes get into the Excise, and in case of being discharg'd for neglect, etc., they turn Poachers.[14]

The deficiencies of the eighteenth-century schools are obvious. The masters were ill paid, receiving less than most lead miners, and their incomes similarly fluctuated with the price of lead, as paying pupils were withdrawn when times were bad. In consequence it is probable that the standards of teaching were very low, even by eighteenth-century standards. 'Children of the lowest classes of the poor are not taught even to read, from the inability of their parents to pay the quarterly sums.'[15] Worst of all was the inadequate number of schools. The total number of children in the schools of Alston parish in 1805 was 209; the population of the parish in 1801 was 4,746. Most children, therefore, never went to school at all. Girls were apparently not taught at any of the Alston schools in 1805. Surviving petitions show, that, by the end of the eighteenth century at least, there was a real desire among the lead miners for their children to be educated. Two petitions survive from the inhabitants of upper Weardale asking for the provision of more schools.

The first, sent to the Crewe trustees in 1778, was from the inhabitants of Killhope and Wellhope:[16]

That your Petitioners as being poor mechanic people labouring in the Lead Mines for our livelihood and many of us having large families of children to provide for, are in general hard put to it to procure for them a Scanty Subsistence, even of the bare necessaries of Life—And indeed if we can, from our hard labour, but get their necessary want of food in some measure supplied, together with a verry scanty Pittance of Cloaths, it is, in general, the stretch of what we can do. As for learning to our dear Offspring, of this (Melancholy to say it) they must really be destitute; as we cannot send them to a distance, and no man will come amongst us for this end, from the small encouragement that we ourselves can afford him; of consequence our Posterity, from one generation to another, must be brought up in a State of Ignorance, not many degrees, removed, in point of religion, above the wild Indians. For how can it be otherwise—We ourselves as soon as we could be of the least service to our Parents in bringing in some little Aid for the family from the Mines, we were set to work therein—our children must follow the same Method—their Offspring must do the same, and so on to future generations.

It may be said in comment that whoever the unknown person responsible for composing this petition might have been, he was far from illiterate himself.

Phrased in a more formal manner, a petition of about 1800 to the Bishop of Durham from some inhabitants of St John's Chapel shows the deficient coverage of charity schools in the enormous parish of Stanhope:

That there being only three free Schools within the Parish, viz. at Frosterley, Stanhope and Westgate and consequently none within 10 Miles of the most distant Parts of the Chapelry where the greater Number of the labouring Poor reside, the Necessity of additional Schools is so strongly impressed upon the Minds of your Petitioners as to preclude the necessity of urging any Arguments in Favour of such Establishments.[17]

In the years 1818 and 1819 educational facilities were enormously improved. The inhabitants wanted more and better schools, and were prepared to help themselves as far as possible. In 1811 a new school building for 200 children to be educated on the Lancastrian pattern was completed in Alston. Funds had been raised by public subscription.[18] But outside help was needed if there were to be adequate schools all over the region. It came from the Church and from the London Lead Company, followed, with less eagerness, by other mining companies.

In 1807, after threatening an action in the Court of Chancery, the Bishop of Durham, Shute Barrington, received a total of £70,000 from the Beaumonts on account of non-payment of royalties due to him as owner of the mineral rights in Weardale. Barrington was a member, and eventually president, of the Society for Bettering the Condition of the Poor, which was primarily concerned with education; he determined to use part of the money for founding schools for the poor in County Durham. In 1809 he founded a school in Bishop Auckland which became a great success. In 1819 he began work in Weardale, the place whence his money had originated. In making this decision on behalf of the Church, he was urged into action by reports of the plans of the (Quaker) London Lead Company to build and finance schools in the areas where its employees lived.

In 1816 the Lead Company had appointed Robert Stagg as its new superintendent of mining works in the north. He was a Quaker himself, apparently unlike his predecessor, and keenly interested in social reform as well as business efficiency.[19] Moreover, the educational experiments of Joseph Lancaster, also a Quaker, were well known by this time. In February 1818 the court of the company passed a resolution agreeing that the 'two schools recommended by Mr. Stagg' should be established.[20] These schools were to be constructed at Nenthead and Middleton, the centres of the company's two chief mining districts. But as the company had been acquiring leases of mines in Weardale since 1800, it was obvious that it would also be interested in education there. That a Quaker company should initiate educational reform in the richest parish in England (judged by the rector's stipend) was not acceptable to the Established Church.

In 1819 Bishop Barrington bought sites for four new schools in Weardale; one at Stanhope, one at Boltsburn in Rookhope, and two in the main Wear valley west of St John's Chapel. He spent £2,000 on the construction of buildings on these sites, and a further £2,000 was invested to provide a regular income for the new schools.[21] The trust funds of the older Weardale schools were amalgamated with the new fund, and these schools as well were supported by it, save for two that were entirely superseded by the new foundations. All the schools were brought under a uniform set of regulations. The teachers were to be members of the Church of England, and the children were to attend church twice on Sunday. The parents of each child were charged 1s 6d a quarter but there were twelve free places at each school. Examinations were to be held in the winter, to enable 'the washer boys or such other children as are absent during the summer months to attend'. One regulation enabled the trustees to close down a school and open another 'in case the precarious nature of mining adventures should at any future period render it necessary to remove the present schools to a part of the dale more convenient to the workmen'.

The two schools established by the London Lead Company were both opened in 1819. From the beginning they were governed by a set of regulations remarkable for their comprehensiveness.[22] The most notable rule was that 'all children belonging to the Lead Company's Workmen, and the Widows of deceased Workmen, shall be required to enter the Company's Week-Day School at Six Years of age ... The Boys to attend from Six to Twelve years of age; and the Girls from Six to Fourteen years of age, if they so long remain under the

paternal roof'. In case of extended absence, every child had to complete
the 'required time of attendance, viz. Six years in case of a Boy, and
Eight years in case of a Girl'. No boy would be employed by the
company 'who cannot produce a certificate from the Company's
Teacher, stating that he has complied with the . . . regulations'.
R. W. Bainbridge, the company's chief agent, was asked by the 1864
commission what happened to a parent who did not obey the regulations. He replied that he would first inflict a fine, and then 'if he still
remained obstinate I should not hesitate to dismiss him from the
service'. No employee had been dismissed for this reason; fines had
been levied occasionally.[23] Parents were charged a shilling a quarter
for each child, and children of parents not employed by the company
were charged 2s 6d a quarter. Any children of employees who lived
too far away to attend the company's schools at Middleton and Nenthead had to attend the nearest school that had been approved by the
company's school inspector, and to attend for the same period of time.
The company would contribute 2s a quarter towards the education of
each child in such cases.

The teachers had at first to be members of the Church of England.
This stipulation was later dropped after a quarrel in 1849 with the
Bishop of Durham. The schools were, however, non-sectarian in
character, each child being required to attend 'twice every Sabbath
such place of religious worship as his or her parents may think proper'.
The masters were each paid a salary of £100 a year, a far higher sum
than was normal both in 1819 and during the remainder of the nineteenth century up to 1870. The method of teaching was, as in all
schools in the district at this time, monitorial, one master instructing
senior pupils, who in turn instructed the junior classes under his supervision. These monitors were also paid small sums.[24] As to the curriculum, Stagg told the Poor Law Commission in 1834 that 'all that is
professed to be taught in the Lead Company's schools is reading,
writing, and so much of arithmetic as suffices to enable the men to
check the overlooker's accounts by keeping accounts of their bargain-
work themselves'.[25]

The new educational projects of the Bishop of Durham and the
London Lead Company stimulated the other mining concerns to
action. In March 1819 the Beaumont chief agent wrote to Mrs
Beaumont that the London Lead Company had publicised its intention
of giving £400 towards the support of Bishop Barrington's new Weardale schools. 'I have considered it my duty to transmit the same for

your information,' he wrote. In February 1820 the agent promised the Bishop's school officer that Colonel and Mrs Beaumont would make an annual donation towards the upkeep of the Weardale schools.[26]

A Greenwich Hospital report stated that on Alston Moor in 1822 the schools 'are increasing in number and improving in management. There are already One Thousand Children under instruction'.[27] The Hospital gave annual contributions for the financial support of all the schools on the Moor. By 1842 it was giving five schools £10 a year each, and a sixth £20 a year. The Hudgill Burn Company endowed the school at Nenthall with £200.[28] The London Lead Company also gave regular sums to schools on Alston Moor and in Teesdale which were attended by children of its employees.[29]

Contemporaneously with the expansion of day schools around 1820 there was a rapid development of Sunday schools. The recently formed Newcastle Sunday School Union sent a delegation to Teesdale and Weardale in 1823 in response to a request for aid in the shape of money and books. In Teesdale there was but one firmly established Sunday school, together with one recently formed by the Methodists in the far west of the inhabited part of the dale, which was 'upon the point of being given up for the want of a room to teach in; that in which they had before taught having been filled with hay'. One new school was established there. In Weardale, which was 'far beyond Teesdale in regard to education and means of grace', three new schools were established, making a total of nine all told, with a total of 152 teachers, as against only fifteen in Teesdale. Alston parish had twelve schools and 180 teachers, and Allendale one school and 25 teachers.[30] The London Lead Company established Sunday schools in association with both its day schools, attendance at which was compulsory for all the pupils at the latter. They were expected to continue attendance at the Sunday schools after they had started working, and were excused only when they had 'passed a satisfactory examination in scriptural knowledge'.[31] By 1842, in Alston parish, 'To every place of worship, I believe without exception, there is attached a Sunday school'.[32]

In 1840 a survey was made of the schools and pupils in Alston parish, giving figures to compare with those already quoted from the Greenwich Hospital report of 1805. There was a total of fifteen schools in 1840, as against five in 1805.[33] There were 702 pupils, compared with 209 at the earlier period. The population of the parish in 1841 was 6,062; in 1801 it had been 4,746.[34]

The 1842 report quotes several statistics and many individual statements about education. A return from the Beaumont concern surveys the educational abilities of the younger employees:

Of 432 boys and young persons under 18 ... 419 are stated to regularly attend public worship; 270 attend Sunday schools; 415 are stated to be able to read an easy book; and 291 have signed their names. Many of these who have not signed their names are still attending to their education in winter, and may yet acquire the art of penmanship.

Dr Mitchell observed of these figures that they showed 'a much more favourable state of things than in other parts of England'. He also found that virtually all miners giving evidence were able to sign their statements.[35]

Outside the London Lead Company's area of operations, the pattern of school attendance was seasonal:

There is always an interval of a few months (generally November to April) every year, during which the boys are unemployed, and at this time they are generally sent to school. They are also in general sent to school at an early age before beginning to work, and the rudiments of education there received are improved, year after year, during the interval of washing.

Twelve was the normal age of beginning work at the washing floors. Most washer boys individually questioned had been to school before commencing work, but only those employed by the London Lead Company had attended for as much as six years. Two or three years was the more normal period. Most of the boys continued part-time schooling in the winter until they were about 15.

In 1849 there began a complete reorganisation of the educational system within the areas controlled by the Beaumonts, the largest of the mining concerns. According to A. F. Foster in 1861, 'these plans were organized immediately after the last great strike among the workmen, and, as I understand, in consequence of the conviction which then forced itself, that their education was necessary as a means of safety for their employees.'[36] In 1849, however, when the Allenheads strike occurred, Thomas Sopwith had been chief agent four years. His diary shows that from the beginning of his tenure of office he possessed a keen interest in education. The strike was a useful excuse for extracting from his employer the necessary money for carrying out his plans. Between 1849 and 1861 seven new school buildings were constructed in Allendale and Weardale, some replacing existing ones, others being completely new foundations.[37] Voluntary contributions were requested from the workmen towards the new schools, but the mining concern provided most of the necessary capital.

All the schools were brought under a uniform set of regulations drawn up by Sopwith.[38] The owner, W. B. Beaumont, retained ultimate control but policy could be recommended by an elected committee—'Two agents, and four miners or other workmen, are to be elected in each district on the first subsistence day in each year, to act as visitors to the schools along with the resident agent of each district.' These individuals could inspect the schools when they wished, entering comments in a visitor's book kept at each school. The whole committee could meet at the request of any two of its members, to recommend any specific action required. Attendance at the schools was open, but not compulsory, to all children of workmen, and others were allowed to attend with the permission of the agent. The fee was sixpence a month per child. The schools were non-sectarian, and precise instructions were laid down concerning the time and content of religious instruction.

About one-third of the regulations were taken up by an outline of the disciplinary system to be employed.

Corporal punishment is disapproved of, but is not entirely prohibited. Every case of its infliction is to be specially reported in writing by the master in the visitor's book. . . . Rewards are given daily, weekly, and monthly by means of tickets, and the punishments chiefly consist in the forfeiture of such tickets as described in the following regulations . . .

Tickets were given for everything—punctuality, good conduct, prompt payment of fees, etc.—and were confiscated in proportion to the magnitude of the offence. The greatest number of tickets an individual could gain in one month was 110. '80 tickets a month for four consecutive months places him in the first class or order of merit, 50 in the second, and less than 50 in the third. Those of the first order receive their books, slates, etc. gratis, those of the second at half price, but those of the third must pay for them in full.'[39] School records were passed on to the mining concern when a boy entered employment.

Sopwith retained his interest in and concern with education all his life and published several pamphlets on the subject. He was always greatly concerned with detail as well as principle. Here is an extract from his diary in 1852 recording a daily routine:

At 9 o'clock I employed, as I often do, the telescope to see the children enter School, which they usually do to a second, and rarely indeed is it that one or two are as many seconds beyond the proper time. One boy was about 3 seconds late ... but it turned out that he had been hindered by an elder boy employed at the Washing Floors, and who is to come to the office on Wednesday to be reprimanded.[40]

Sopwith knew precisely what he wanted the schools to do:

I do not think it either practicable or useful to aim at any high cultivation of mental energy [on the part of the miners] beyond those which can be directed to the advancement of their health, the management of their ordinary matters of business, and the desire to occupy their leisure time in temperant and innocent occupations. I would teach them punctuality, prompt payment, cleanliness, observance of the rules of health, moderation and aptitude in drawing.[41]

In 1861 Foster compared the London Lead Company and Beaumont schools. Both had an elaborate system for regulating the children and linking their educational records with their working lives. Both compared very well indeed with the Government schools inspected—the identical mechanical routines of the latter contrasting greatly with the individual idiosyncrasies of each of the lead mining schools, where a good teacher was given far greater freedom in method and curriculum. Educationally the London Lead Company schools were 'decidedly superior' to the Beaumont schools in 'more intellectual' subjects, but the latter were better in 'singing, drawing, and other matters of taste'.[42] But the great distinguishing point of superiority of the London Lead Company's system was that schooling was strictly compulsory. In Middleton in Teesdale, Foster found the best educational conditions in the whole of the north-eastern region he had inspected.[43]

Thus by the mid-nineteenth century the two largest lead mining concerns provided extensive educational facilities for their workmen. Some smaller companies had similar schemes, the Rodderup Fell Company in 1864 automatically deducted 6s per child per year from a parent's wages to pay for education. The schools were provided for genuinely charitable reasons, reinforced by self-interest, as they provided a means of disciplining future workmen in their impressionable years. The standard of education was higher than in most other areas of England. Foster's report of 1861 quotes statistics of the reading and writing abilities of *parents* in the coal mining areas of Durham and Auckland unions, and of the lead mining areas of Allendale, Weardale and Derwentdale:[44]

Coal mining area	Fathers (%)	Mothers (%)
Can read	78	72
Can write also	64	47
Can neither read nor write	22	28
Lead mining area		
Can read	95·7	90
Can write also	90·4	82
Can neither read nor write	4·3	10

These statistics are a striking demonstration of the superior educational environment of the lead dales in the mid-Victorian period. The 1870 education Act coincided with the final decline of the lead mining industry and the gradual migration of a large proportion of the inhabitants of the dales. The numbers of schools contracted instead of expanding after this date.

Libraries

Parallel with the growth of the number and quality of schools in the area, there developed other opportunities for self-education and intellectual recreation in the form of libraries. By the end of the eighteenth century libraries did exist in the mining region. At Westgate, in Weardale, a subscription library was founded in 1798 which lasted at least until the end of the nineteenth century. There were probably others of which no record remains.[45] One primitive form of library, gathered together by co-operative effort, was the book club: 'A few join in contributing two-pence or three-pence per week for the purchase of books, which are lent out to subscribers; and when the club breaks up, the books are divided by lot.'[46] Clubs of this nature were still in existence in 1842.[47]

After about 1820 there was a great expansion in the number of libraries. In a Greenwich Hospital report of 1821, the foundation of a library in Alston itself was recommended, as 'there is still much want of a parochial library in the village'. The Hospital agreed to give this project financial support, and by 1842 there were no fewer than four libraries in the town of Alston with three others elsewhere in the parish.[6] One of them was in the new Mechanics' Institute. In Weardale two more subscription libraries opened between 1820 and 1840, besides the one at Westgate. Bishop Barrington contributed £50 to the support of these libraries.[47] The Sunday schools generally had a collection of books for the children to take out.[48]

But the greatest contribution towards the provision of libraries in the area came from the mining companies. The leader in this field was, as usual, the London Lead Company. Coinciding with the foundation of the schools at Nenthead and Middleton, the company bought books to form libraries in them. From time to time later in the century money was donated towards a further supply of books.[49] The company also aided the subscription libraries. By 1850 it was subscribing annually to sixteen libraries in its area of operation. Those at Nenthead and Middleton lent books without charge to all workmen and their families.[31] In addition, at these places reading rooms were built as social centres for the men: 'The members attending the room contribute 3d monthly towards the purchase of newspapers and periodicals; an exception being made to youths under 18, who only contribute 2d monthly'.[50]

The Beaumont concern did not provide libraries for its work people until after the arrival of Thomas Sopwith as chief agent. He founded four libraries in Allendale and Weardale, three in 1848 and one in 1850. Money was given to provide a building and furniture at each place, the library being conceived as a club rather than as a mere collection of books. Members had to pay an annual subscription of a shilling a quarter for the first four years of membership, subsequently reduced to a shilling a year.[51] Sopwith noted with some pride in his diary that he had established alongside these adult libraries 'what I have called Children's Libraries . . . the object of which is to afford young children a good selection and frequent change of amusing books.'[52] The children were not charged for the use of the books. In Weardale the library was associated with another Sopwith foundation, the Weardale Miners' Improvement Society, which combined 'the principles of a temperance society and a night school with those of a Mechanics' institute and public library'.[53] Both here and in the ordinary Beaumont libraries in Allendale there were programmes of lectures—Mr Sopwith himself being the most frequent lecturer on all subjects from astronomy to the history of Egypt.[54]

As to the numbers and contents of the actual books provided, little evidence survives. In a survey of the number of volumes in selected libraries printed in Foster's 1861 report, the largest library was the old subscription library at Westgate, with 1,387 volumes; Middleton had 1,189; Allenheads, 1,177; Nenthead, 937. Stanhope parochial library had only 150 volumes, while another Weardale library listed in the 1851 educational census had only twelve volumes![55] These figures are not very high, and membership of the libraries was apparently taken up only by a small proportion of the workmen and their families. Foster lists the number of members in each of the two London Lead Company libraries, and the three Beaumont libraries in Allendale. The highest was at Middleton, with 139 members; the lowest at Allenheads, with 102. The books were probably mainly of the type intended to convey 'moral and other useful knowledge'. No catalogues appear to survive. Sopwith noted in his diary, however, that books were explicitly bought for the Beaumont libraries to provide recreation as well as instruction.[56]

Victorian visitors to the lead mining region eulogised the miners' intellectual leanings. Here was an apparent rural area, its population scattered over many square miles, which possessed schools and libraries as good as those in a closely knit urban community—yet without too many of the physical disfigurements of industrialisation. The rosiness of the picture was undoubtedly exaggerated. Sopwith recorded in his book *Alston Moor* of 1833 the surprise of a friend on hearing a miner working at the face quote Blackstone's *Commentaries* 'both with accuracy and direct reference to the subject of discussion'.[57] The story may be true, though it is hard to believe. But an undoubted example of unusual intellectual activity by those isolated miners was the publication of a book in 1851 entitled *The poetic treasury: being select pieces of poetry from one hundred different authors . . . designed for the benefit of the working classes.* The preface, inscribed 'West Allendale, June 18th 1851', expresses the wish that the work 'may be of some service to members of the working classes wishing to read at the end of the day'. It would be interesting to know how many of their fellow miners bought it.

NOTES

[1] 1842 report (Mitchell), p. 754.
[2] *State of popular education in England.* Royal Commission report, 1861 (vol. 2), p. 323.
[3] Eden, vol. 2, 1797, p. 170.
[4] Greenwich Hospital report, 1821.
[5] Sopwith, 1833, p. 128.
[6] 1842 report (Mitchell), p. 753.
[7] Ritchell, p. 19.
[8] Based on the reports of the Charity Commission, vol. 21, 1829, and vol.

23, 1830. For an extended account of Weardale charity schools, see J. L. Dobson, 'Charitable education in Weardale', *Durham Research Review*, 2, 1955–59, pp. 40–53.

⁹ Dobson, pp. 47–8.

¹⁰ Greenwich Hospital report, 1805, pp. 202–6.

¹¹ Ritchell, p. 17.

¹² Minutes of the Committee of Council on Education, 1840–41, p. 151.

¹³ Adm. 66–97, 9 June 1780.

¹⁴ B/B 50, 4 January 1796.

¹⁵ Greenwich Hospital report, 1805, p. 205.

¹⁶ Printed by Dobson, p. 44.

¹⁷ Thomas Bell MSS.

¹⁸ F. Jollie, *Cumberland Guide*, 1811, appendix, XXV.

¹⁹ For an account of the London Lead Company's educational policies, see Raistrick, 1938, pp. 58–76.

²⁰ L.L.C. 18, 19 February 1818.

²¹ For the founding of the Barrington schools in Weardale, see Dobson, pp. 46–50; report of the Charity Commission, vol. 21, 1829; W. Whellan, *History of Durham*, 1856.

²² Printed by Raistrick, 1938.

²³ 1864 report (vol. 2), p. 376.

²⁴ *State of popular education in England*. Royal Commission report, 1861 (vol. 2), p. 367.

²⁵ Poor Law Commission report, 1834, appendix A, p. 139ᴀ.

²⁶ B/B 51, 6 March 1819 and 5 February 1820.

²⁷ Greenwich Hospital report, 1822, p. 22.

²⁸ 1842 report (Leifchild), p. 686.

²⁹ 1864 report (appendix), p. 18; Raistrick, 1938, p. 60.

³⁰ Newcastle Sunday School Union report, 1823, pp. 20–2, 38–9, 47.

³¹ 1842 report (Mitchell), p. 750.

³² Ibid, p. 768.

³³ The 1840 figure included dame schools, which may not have been included in the 1805 figures. Excluding the dame schools, there were nine schools and 532 pupils.

³⁴ Minutes of the Committee and Council on Education, 1840–41, pp. 148–51.

³⁵ 1842 report (Mitchell), p. 752.

³⁶ *State of popular education in England*. Royal Commission report, 1861 (vol. 2), p. 347.

³⁷ Ibid, p. 345.

³⁸ Ibid, pp. 387–92.

³⁹ Ibid, p. 346.

⁴⁰ Sopwith, *Diary*, 29 March 1852.

[41] Ibid, 27 January 1858.

[42] *State of popular education in England.* Royal Commission report, 1861 (vol. 2), p. 370.

[43] Ibid, p. 373.

[44] Ibid, pp. 332–3.

[45] There was a printer resident in Alston by the 1790's. A periodical entitled *The Alston Miscellany* commenced publication in 1799 and ran until the end of 1800.

[46] Mackenzie, vol. 1, p. 207.

[47] 1842 report (Mitchell), p. 757.

[48] Ibid, p. 769.

[49] Raistrick, 1938, pp. 70–1.

[50] 1864 report (appendix), p. 18.

[51] Book plate of Allen Mill library.

[52] Sopwith, *Diary*, 1 July 1865.

[53] *State of popular education in England.* Royal Commission report, 1861 (vol. 2), p. 355.

[54] In an interesting note in his *Diary* for 1849, Sopwith recorded that 'Mr William Chambers [the Edinburgh educational publisher] and I had it under consideration to publish a series of pamphlets under the name of *Allenheads Tracts* but this intention was not carried out. The general policy at the W.B. lead mines . . . has been to aim at privacy and to avoid publicity in the eyes of the world.'

[55] *State of popular education in England.* Royal Commission report, 1861 (vol. 2), p. 396.

[56] Sopwith, *Diary*, 5 June 1860.

[57] Sopwith, 1833, p. 139.

XIII

Conclusion

In the eighteenth and nineteenth centuries the lead industry passed through a social as well as an industrial revolution. Superficially, life in the mining dales changed less between 1750 and 1850 than in the neighbouring coalfields and shipyards. The wild and romantic Pennine scenery remained little corrupted by industrialisation. Mining continued to be governed by apparently the same elaborate system of sub-contract. But population increased by a factor of at least three, and below the surface (metaphorically speaking) social institutions changed fundamentally.

Technological advances in underground haulage and ore dressing at the beginning of the nineteenth century forced organisational changes on the mine owners. In these fields sub-contracting was either abolished (the London Lead Company) or regulated so closely that the sub-contractors were direct employees in all but name (the Beaumonts and the Derwent Mining Company). The contracts governing actual ore getting became tighter, reducing the practical status of the theoretically independent bargain taker to that of employee. The proportion of agents to workmen increased, allowing greater supervision. The miners were paid more regularly—and were expected to work more regularly. Outside working hours, little of their social life was not influenced by the owners by 1850. Education, churches and chapels, benefit societies, even organised amusements were provided or subsidised by the mining companies. But misdemeanours in private life—drunkenness, fathering a bastard child, etc.—were as much the province of managerial discipline as any offence during working hours. At work and away from it, the lead miner was dominated by his employer.

The employers' object was to increase profits. Benevolence was one motive behind the provision of welfare services, but it was scarcely altruistic. Efficient management meant controlling the labour force at home as well as at work, and educating the next generation into a proper sense of discipline. The London Lead Company, the most efficient employer, introduced the changes earliest; the other large companies gradually followed suit. The miners wanted higher and more secure remuneration for their labours. In exchange for greater security, they were prepared to surrender many of their traditional freedoms.

The emergence of a new management ethos occurred simultaneously with two other important social changes. The first was the growth of Methodism, the converts to which were more prepared to accept puritanical regulation. The second was the weakening of the miners' bargaining position *vis-à-vis* their employers. In the eighteenth century the mining industry, despite occasional depressions, was expanding fast, and employment kept pace with population growth. In the nineteenth century, expansion slowed down but population growth did not. Also, because of the increased liquidity of the money market, the nineteenth-century mining employer, unlike his eighteenth-century predecessor, was rarely in debt to his employees. These factors were important, possibly vital, to the extension of managerial authority.

All the changes that did occur within the mining industry took place within the bargain system. Only the Derwent company abandoned this completely—and replaced it with a Cornish variant. Except in this case there was never a major influx either of new management or of new labour from outside the region. Masters and men were conservative, and used to traditional forms. Revolutionary change was disguised by old convention.

It has proved impossible to measure quantitatively the changes in the standard of living over the period. Qualitative judgments, however, can be made. The lead miner of 1850 received his wages more regularly and securely than his predecessor of 1750; most of his chances of making a modest fortune by a rich strike had gone. He was in less danger of near or actual starvation in 1850, but most of his feasts and festivities had disappeared. Life was more secure, but also more regulated; 'Big Brother' was very close. In the 1870's and 1880's, with the collapse of the industry, the security disappeared as well.

Appendix 1

London Lead Company court resolutions laying down general rules for the administration of mines and estates

From L.L.C. 13A and 17

6 October 1785

The Court came to the following Resolutions:

That the Company's Pays in the North which used to be paid Half-yearly be in future paid only once every Year up to Michaelmas 1785 and so to be continued; and that the following instructions be sent to their several Agents there.

That the following regulations be adopted and carried into execution as soon as can be done consistently.

Viz that the full pays be made once a Year at Michaelmas and that no One be paid for but what is Washed and Weighed up, and none sent but with printed tickets of the quantity from the Mine to the Mill.

That no Agent be permitted to deal in any Commodity made use of by the Miners or Smelters; nor that they be permitted to be concern'd underhandedly with any person who shall deal in such goods: nor be concerned in Whimseys or Letting Horses to draw Waggons, or Wood, or any thing of that kind for the Company: or have any allowance of Coals or Candles.

That the monthly advance money & also what becomes due to them be paid to them themselves & no other person, in money and not in any kind of Bank Notes, nor any Trades people attend at the pay: nor to any order written or otherwise, except the Miner is ill & cannot attend himself.

6 December 1810

The Court have unanimously agreed to the following resolutions...

4th. That a Fund be established for the Relief of maimed & decayed Workmen employed by the Company upon such Plan and under such Regulations as shall be agreed upon . . .

8th. That no Gunpowder be used by the Company's Workmen but what is furnished by Order of the Court & to be charged to them at prime Cost, with only the addition of Freight & Carriage.

9th. That the Candles be purchased upon the best Terms, paid for Quarterly and charged in the same Manner as the Gunpowder.

10th. That in letting of Bargains no person whatever shall have an hireling employed on any account in the Company's Works, except a Workman who may fall sick and has taken a Bargain previous to his sickness; and he shall only be allowed this till that Bargain be finished.

11th. That no Bargain be let for a longer Term than a Quarter of a Year.

12th. That as few Wagemen as possible be employed, but that in every instance where it can be accomplished, the work to be done by the Peice, of what nature soever.

13th. That for engaging of Whimsey or Level Horses used by the Company, Notice to be given to the parties letting the same for receiving proposals for yearly Contract; the proposals to be given under seal and forwarded to the Superintendent.

14th. That all Stores such as Wood, Rope, Iron etc. necessary for the Company's Mines, shall be purchased of the first wholesale houses at Newcastle at the short Credit, & the benefit of the discount taken . . .

17th. That the Storekeeper shall keep a correct Account of . . . Wood, Rope, Iron, Gunpowder, Candles and other Stores; keeping a similar account of the same, making a distinct Account of what is to be reckoned for with the Workmen from what is paid for by the Company . . .

20th. That no Agent, Clerk or Assistant, or any party on their behalf shall have any concern, directly or indirectly, in selling Gunpowder, Candles, or any other material used in the Company's Mines, or be concerned in the Corn Trade, or in the sale of Shop Goods to the Company's Workmen, or any others, either directly or indirectly.

21st. That no Agent, Clerk, or Assistant, or any party on their behalf, shall have any Mine or Shares of Mine, or be concerned either in Mine adventuring, or in the purchase of Ore, directly or indirectly, or in any shape whatever . . .

22nd. Coals and Candles are not allowed to any Agent or otherwise.

7 March 1812

25. That no Agent, Clerk or Assistant or any party on their behalf shall have any Mine or Shares of Mine or be concern'd in Mine adventuring or in the purchase of Ore, directly or indirectly or in any Shape whatever or be concern'd in the Farming Business or employment except keeping a Horse or Cow or two for the use of their Families . . .

28. That every charge against the Workmen for Drawing Work, Tools, Crushing Mills etc. as well as advances, Candles & Gunpowder, to be passed thro' the Due papers & no part to be reckoned or taken off them in any other way whatever . . .

Appendix 2

Arrangements recommended to Colonel and Mrs Beaumont for the future agency and management of their lead mines

From Blackett of Wylam MSS, Northumberland County Record Office

At Weardale

Mr. Emerson	Principal Agent	£100 salary
Mr. Geo. Crawhall	Assistant to Do.	£60 „
Mr. Josh. Harrison	Clerk or Bookkeeper to Do.	£50 „
	Inspector of Miners at Breconsike	£60 „
	Do. of Miners at distant places.	£60 „
Martin	Do. of washing ore at Breconsike.	£50 „
	Do. of Do. at distant places.	£50 „

Joseph Potts and John Foster to be discontinued—also George Dixon except he be thought a proper person to be inspector of the Miners at distant places.

At Allenheads

Mr. Crawhall	Principal Agent	£100 salary
Mr. Wm. Crawhall	Clerk or Bookkeeper to Do.	£50 „
	Inspector of Miners	£60 „
	Do. of washing ore (the same Person as at Colecleugh	£25 „

Mr. John Crawhall to be discontinued.

At Colecleugh

Mr. Joseph Dickinson	Principal Agent	£60 „
until he be succeeded by Mr. Thos. Dickinson.		
Mr. Alec Prothero	Clerk or Bookkeeper.	£50 „

| Mr. Thos. Dickinson | Inspector of Miners until he succeeds Mr. Josh Dickinson. | } | £60 salary |
| | Inspector of washing ore, the same person as at Allenheads. | } | £25 ,, |

Westgarth Forster and Josh. Dickinson Junr. to be discontinued.

It being the practice at Allenheads to pay £10 per ton for Tontail ore the washers finding Fuel—or 9–10–0 per ton, when Coll. Beaumont finds it, we recommend in future, that it be made a General Rule for the Coll. to find the fuel at all times, likewise that in all future bargains with the workmen no Shifts be allowed at any of the Mines (which has been the practice at some of them) but that all the Deads be — removed and cleared away at the Expence of the contracting workmen — so that Coll. Beaumont may not be put to any Costs in removing Deads except in the revivial of old workings.

The Principal Agent at each of the three Mines to have the privilege of employing Horses to draw all the Lead Ore, Bouse Ore, Dead work &c from their respective Mines, also for the Carriage of Timber or other Materials, at such rate or price per Shift or otherwise as may from time to time be deemed fair and reasonable, taking into consideration the nature of the work, the price of Hay and Corn and every other circumstance attending it, but such Agent not to have or receive any other privilege advantage or perquisite whatsoever.

No Agent, Clerk, Inspector or other person to have the privilege of nominating a Hireling or Wageman, so as to derive a benefit from any Bargain or contract for raising ore, driving levels or otherwise but such persons only who may contract for that purpose and — whose names may be inserted in the Bargain Book at the time of contracting, to be intitled to the benefit or advantage arising from such Bargains.

A general Inspector is also recommended, whose duty it will be to examine the state of the different Mines in the course of working previous to the expiration of such Bargain, so as to determine the proper price to be given at each letting, which letting he must attend — he must also point out all such neglect, or other observations may occur to him on such inspection — likewise point out such places of Trial for Ore as he may think advisable and view such other places as may be recommended by the Principal Agent, whose duty it will be to point out to him every favourable and encouraging prospect of Trial upon which he must state his opinion in writing, along with every other matter or thing that he may from time to time recommend to be done, to the principal Agent at each mine and also to the Chief Agent C. Blackett Esqr. at Newcastle.

Duty of Principal Agent. Altho in all well regulated Systems every department should be appropriated to a proper Agent so as that in case of any
neglect it may be known on whom the blame should fall, yet such principal
Agent must in the present instance take upon himself the *whole responsibility*,
consequently consider himself the inspector of *every department* within his
district, and if any of the assistants neglect to perform their duty, he must
acquaint Mr Blackett the Chief Agent therewith, but not discharge or
appoint any assistant Clerk or Inspector without the direction of such Chief
Agent — He must also apply to the Chief Agent upon all occasions when
anything material occurs or is wanting, and observe and execute such orders
and directions as he may from time to time receive from him or the general
Inspector of the Mines, to whom he must point out from time to time such
fresh or neglected places as he may think deserving of a Trial for Ore — He
must also prepare a plan delineating the present state of the workings at each
Mine in his district, in order that the Inspector of Mines may continue the
description of the workings on such plan according to the directions given to
him.

An Assistant to the Agent at Weardale is also recommended on account of
the Mines there being so extensive and at so great a distance from each other,
as to render the undertaking too much for one Agent to pay proper attention
to, it will therefore be the duty of such Assistant, to employ his time, in viewing the Mines, Dialling, laying down plans from the Minute Books of the
inspecting Miners, keeping Accounts, and doing all such other business as his
Principal may from time to time direct.

The Duty of Clerk or Bookkeeper will be to keep all Accounts necessary at each
Mine, and which must be done in the *precise form and method* that has hitherto
been practised — he must also assist in laying down plans and such other
employment as the principal agent may direct.

The Duty of the Inspector of Miners. Every such Inspector is proposed to be
selected from the working Miners at one of those Mines over which he is not
to be employed as Inspector — He must be capable of taking levels, hatching,
and laying down correct plans of the different workings, not merely describing
the progress of such workings, but also the dead work done at each place. All
which must from time to time be regularly inserted in a plan kept for that
purpose in order to shew the progress and state of the workings, which plan
the principal Agent must provide and describe and lay down thereon all the
workings, as well as all the Drifts and Levels carried on to the present time,
and upon which plan so prepared the Inspector of Miners or whom the Agent
may appoint must insert at the end of every week (not delaying longer) the
progress made in the workings and dead work — and to the intent that recourse may at all times be had to the Memorandum or Minute Book of each

such inspector of Miners, a fair Copy thereof must be made and deposited in the office.

It will be required that such Inspector go into all the Mines within his district every day, and take particular care that all the working Miners do their duty — that they are properly supplied with Timber, Ropes &c and that no improper or extravagant use be made of any kind of materials necessary for the works, or any wilful waste committed — that all the underground Roads be kept in good order — that each Mine be kept clear, open, and free from Bouse, as well as all such Deads as are not wanted to pile up in proper places to support the Roof and save the use of Timber.

The Inspector of Miners from the nature of his Business must know, when, where, and for what purposes Horses will be necessary to be employed, he must therefore give directions accordingly to the principal Agent who supplies them, and must also attest all such Bills or demands as such Agent may have upon Coll. Beaumont for any work done by his Horses — He must likewise communicate with the principal Agent from time to time, and give him sufficient notice to prepare and provide materials of all kinds that may be wanted, and inform him of all complaints, neglects, or any other matter of which he does not approve, in order that such grievances may be removed, in the manner that such principal Agent may think proper, and in general he must apply to and follow his directions on all occasions.

The Duty of the Inspecting Washer. This person should also be selected from the most respectable of the Men actually employed in that branch, and must be fully acquainted with all the various modes of washing ore — He should be confined to such an extent, as will allow him to attend the People employed, twice or thrice a day, and particularly at the time the carriers take up the Ore, when, any that may have been improperly washed and covered over with clean ore as a deception will be exposed — It will be his duty to take care that all the ore be clean and perfectly washed, and not to permit any to be taken away to the Smelt Mills that is otherwise — Also to keep an Account of all the ore so to be taken to those Mills, and deliver to the Carrier a note to go along with it, specifying the quantity from each place — He must also frequently compare his Account of Ore delivered with the Account kept by the person who receives the same at the different Smelt Mills.

The Receiver of the Ore at the Smelt Mill should not only see that the quantity received corresponds with that specified in the note from the inspecting washer, but if he finds any which he thinks imperfectly washed, he must signify the same, upon the note of the Carrier who brings it, in order to shew the work from which it is brought, of which he must give immediate notice to the principal Agent at that work.

The different Smelt Mills and refinerys are presumed to be in such an approved state of management, as not to require any immediate investigation or alteration except as to the receiver of the Ore, as before observed, and an addition to the Salary of the Agents employed at those places which we beg leave to reccommend may be increased to the undermentioned Sums vizt.

At Blaydon Refinery — Mr Robert Mulcaster to 120£

                     John Mulcaster to 50£

Dukesfield   Mr Westgarth to 100£

Allen Mill   Mr Wm. Dixon to 50£

Rookhope   Mr Thos. Smith to 60£

19th January 1805   *James Cockshutt*

                                                 Chas Bowns.

Sir,

     We have taken the above written arrangements into consideration and do approve the same, therefore request that you will give the necessary directions for having them carried into full effect, also that you will supply the vacancies and make the removals therein pointed out as soon as can conveniently and with propriety be done, likewise that you will inform the several and respective Agents of the duty stated in such Arrangements and which will be required of them in future and of the Augmentations made to their Salaries.

To C. Blackett Esq.   *Tho Rd Beaumont.*

    Chief Agent at   *Diana Beaumont.*

    Newcastle.

London, Jan. 23rd 1805.

Appendix 3

Petition from the Weardale miners, 1796

From Hexham manorial papers, Newcastle University Library

St John's Chapple Weardale Augst. 16th.

To the Hon Col. Beaumont M.P.

The humble Petition of the Commetee of Weardale Miners

Sheweth/ That your Petitioners did in Novm. last forward a Petition to you which they believe was intercepted by C. Blackett Esq. Newcastle that your Petitioners have boath before and since that time received various insults from your agent at Newhouse and weare we to inumerate all the advantages resulting from the workes to the agent and his friends you would realy be astonished; suffice it to say that the Money he receives for one horse employed in drawing the Ore and Strata amounts to the enormous sum of 120£ per annm and the number of horse's imployed in the mines at one third less price would aford comfortable livings to Several poor famleys of which a great many have no employment at present. And not withstanding the exorbitant price of Candles and Gunpowder and other Untencels and your petitioners have Candles of an inferiore Quality and on account of the long carriage they are much brok which renders them worse by 2/6d per Doz. than those of which other miners use and we are even bereft of the wrappers in which the Barrels are contain'd for the safety of the Powder and Conveniences of carriage which wrappers are delivered in barter at 8d per Stone to sellers of earthenware. That your petitioners are very assiduous workmen and not envious of any good to any Man, but at the present Juncture are utterly incapable of supporting themselves and famley's by reason of the low prices given them for working in the lead mines, and when we turn Our eyes to the envious prospect of labours our Breathren in business receiving an advance of prices from the London Co. so fare as 35s to 40s and upwards per. Bing, while ours remains Parrallel with the times when the nessaries of Life weare not half the value they are now. But at the same time a certainty remains in us that all this is unknown to you wose wisdom and goodness fille us with hopes of relief. They theirfore humbly pray that you will interpose your authority in their behalf and order them Sufficient wages for their work.— And one guini per month each man for their present support, a pay three months sooner than the wonted time and such treatment as Rational beings are entiteled to, which they have so long, so often but in vain expect'd from the agent here with such regulations as you may think expedient. Then shall the name of Col. Beaumont be recorded in the annals of futurity as the Saviour of the oppressed and dwindling Country. The weeping mothers shall suppress their tears and teach their lisping ofspring to bless the preserver of their lives, while they themselves invoke the Almighty to shower down innum. Blessings on the head of their Benefactor and long containe him in that exalted station for which Heaven has selected him to defeat the cruel Designs of unfeeling Individuals.

And your Petitioners as in duty bound shall ever Pray.

Edward Byeres   *John Robson*   *Thos. Dixon.*
  and 181 others.

Appendix 4

Petitions from the Weardale miners, September 1818

From Blackett/Beaumont records

To Martin Harrison Esq.

The humble petition of the Miners in Weardale—Sheweth that your petitioners are suffering and have long suffered the greatest distress owing to the pressure of the times and the low price given for raising Ore—this they have endured with great patience and fortitude in expectation of better times—We have continued working a long time confiding in Lady Beaumont's promise that our Wage shall be advanced, as soon as the price of Lead advanced. Lead has consistently advanced since that period, but we have only received five shillings more per bing, and that only for the poorest parts of the mine, great numbers of us are not making our Subsistence Money, & have been under the necessity of seeking relief, upwards of 400 of us (including their Families) are on the Parish. Mr. Crawhall says that he has not power to give more at present, our strength and spirits are gone, numbers of us nearly without food and raiment suffering extreme poverty.—We feel ourselves obliged from these distressing circumstances to humbly request that you will have the goodness to advance the price for raising Ore not less than 10/– per Bing and other work in proportion; in order to enable Men to make 15/– or 16/– per week, and to advance our Subsistence Money to 40/– per Month—from your wonted goodness & generosity we humbly hope you will grant our reasonable request & your petitioners will ever pray &c.

To W. Beaumont Esq. M.P.

Honourable Sir

We the Miners in Weardale having long laboured under the greatest distress, have felt ourselves obliged to petition for an advance of Wages. We find that our present subsisting Money which is only 7/6 a Week, much too little to purchase the necessaries of life. We have therefore petitioned for 10/– a week. Our requests have not yet been granted. Mr. Harrison has made us an offer of 5/– per Bing more than we have had, but it is too little, as those places which wo'd have the 5/– advance cannot be worked under 10/– to enable Men to make a sufficiency to support themselves and families, and unless the Subsisting Money be advanced to 10/– a Week, it will be impossible for the greatest number of us to get the necessaries of life, as our Credit is entirely gone—We have taken the liberty of enclosing the petition, which we sent to Mr. Harrison and humbly beg the favour of you to interfere in our behalf—We hope that your influence will prevail on your worthy Mother, to grant us our requests,—We feel fully assured from your known generosity & benevolence, that you will do whatever is in your power, to render our situations more comfortable than they have been.

Appendix 5

Rules and regulations to be observed and kept by the contractors for washing bouse and wastes at Barney Craig and Coalcleugh lead mines entered into this 23rd day of April 1833

From Blackett/Beaumont records

1. We the undersigned Contractors for Washing Bouse and Wastes do agree to give the Boys the wages specified in the returns made at the Office for working the following Hours viz: each Boy to commence working at 7 o'clock in the morning and continue to 12, to be allowed one Hour for dinner and commence work at one o'clock and continue till 7 at night, in case any should neglect working the above Hours, to forfeit so much of their day's labour as hereby set forth, if an Hour behind time in the Morning to have a quarter of the day's wages deducted or work two Hours overwork, and if two Hours behind the specified time, not to be allowed to commence till 12 o'clock with $\frac{1}{2}$ a day's work (at the approbation of the Contractors) and the same forfeits to be exacted in the same proportion against defaulters at any period of the day, this to be in force for every working day in the week except Saturdays when the Boys will be allowed their shifts for working from 7 to 12 in the Morning while the days admit of working the above specified time, and in the Autumn when the days will not admit of working so late, the Boys to work on the afternoon of the Saturdays at the discretion of the contractors to make up the deficiences.

2. A Person will be appointed to Ring the Bell at the above precise times when every Masterman washer is expected to be on the spot and see the Boys are in readiness to commence, all defaulters to be reported to the Ore inspector. The Contractors to make a return of the exact time they are employed together with their Boys with each separate heap of Bouse to the Ore inspector, in default of this compliance the contractors will no longer be considered workmen of Mr. B.

3. It has been a regular practice with the Barneycraig washers to knock a considerable part of the pickings taken from the Grate which cost them $\frac{3}{4}$d a Kibble, we can have it taken to the Crushing Mill and Grinded at $1\frac{1}{2}$d per

Waggon and can save the sludge or slime Ore equally well, it being exposed to a similar stream of water in both operations, the Ore from the Crushing Mill is generally better dress'd particularly the round part or sieve Ore it being freed more from ends or chats by being more regularly sized and I could recommend as much of the poorer Bouse as possible to be taken direct to the Crushing Mill it being washed at less expence and also improves the quality of the Ore very much.

4. The Miners not to be allowed to render the Washers any assistance in the Washing of their own Bouse, except in two or three instances with the top sill work. The Miners generally complain of the Washers dressing their Ore finer than necessary which makes it unpleasant for the washers themselves as well as the Ore Inspector, and that every facility be afforded in getting the Bouse regularly brought to the surface to keep the washers in employment.

5. Henry Bell & Partner to have 2/3 per Bing for all the Breaking sent to the Crushing Mill after being sufficiently wash'd.

6. Each Boy to have 10/– per Month subsistence Money advanced so long as his wages does not exceed 5/– per Week, from 5/– to 8/– per Week 15/– per month—from 8/– per Week upwards 20/– per Month provided the contractors certify they have the same earned.

Appendix 6

Numbers employed at Alston Moor lead mines

Note: the quarters are numbered thus:

1 Christmas (of previous year) to Lady Day.
2 Lady Day to midsummer.
3 Midsummer to Michaelmas.
4 Michaelmas to Christmas.

The figures are broken down into pickmen, labourers and washers. Only specimen years are given in full; for other years, only the total numbers employed are given.

(a) *Persons employed each quarter, 1738–67*

Year	Quarter	Total	Year	Quarter	Total
1738	1	–	1742	3	296
	2	281		4	309
	3	297	1743	1	260
	4	305		2	229
1739	1	315		3	261
	2	416		4	247
	3	348	1744	1	226
	4	350		2	205
1740	1	308		3	180
	2	360		4	187
	3	354	1745	1	156
	4	373		2	215
1741	1	334		3	267
	2	340		4	275
	3	345	1746	1	298
	4	357		2	352
1742	1	305		3	355
	2	294		4	355

Year	Quarter	Total	Year	Quarter	Total
1747	1	395	1752	3	708
	2	411		4	670
	3	444	1753	1	809
	4	423		2	803
1748	1	445		3	858
	2	427		4	866
	3	596	1754	1	–
	4	605		2	875
1749	1	647		3	867
	2	676		4	856
	3	668	1755	1	804
	4	650		2	763
1750	1	640		3	748
	2	646		4	824
	3	629	1756	1	829
	4	634		2	827
1751	1	636		3	856
	2	680		4	852
	3	630	1757	1	827
	4	608		2	889
1752	1	666		3	918
	2	705		4	986

Year	Quarter	Pickmen	Labourers	Washers	Total
1758	1	666	144	88	898
	2	673	117	88	878
	3	693	120	90	903
	4	647	109	110	866
1759	1	659	121	100	880
	2	607	104	90	801
	3	605	106	76	787
	4	626	100	66	792
1760	1	651	94	65	810
	2	573	88	67	728
	3	584	85	62	731
	4	653	100	98	851
1761	1	693	110	96	899
	2	698	114	100	912
	3	707	99	93	899
	4	783	124	118	1,025

Year	Quarter	Pickmen	Labourers	Washers	Total
1762	1	830	128	132	1,090
	2	744	124	93	961
	3	848	124	152	1,094
	4	820	130	156	1,106
1763	1	925	118	142	1,185
	2	954	143	136	1,233
	3	893	139	186	1,218
	4	938	143	169	1,250
1764	1	915	165	145	1,225
	2	894	168	182	1,244
	3	898	167	191	1,256
	4	870	171	185	1,226
1765	1	958	171	130	1,259
	2	1,301?	151	209	1,661?
	3	879	178	222	1,279
	4	887	183	222	1,292
1766	1	857	237	216	1,310
	2	914	210	257	1,381
	3	962	215	275	1,452
	4	1,013	234	260	1,487
1767	1	1,076	264	264	1,604

(b) *Persons employed each quarter, 1818–44*

Year	Quarter	Total	Year	Quarter	Total
1818	4	974	1822	3	1,419
1819	1	1,008		4	1,415
	2	1,254	1823	1	1,161
	3	1,239		2	1,556
	4	1,082		3	1,577
1820	1	1,190		4	1,469
	2	1,387	1824	1	1,419
	3	1,283		2	1,575
	4	1,056		3	1,570
1821	1	1,253		4	1,458
	2	1,347	1825	1	1,375
	3	1,360		2	1,572
	4	1,343		3	1,578
1822	1	1,179		4	1,573
	2	1,418	1826	1	1,260

Year	Quarter	Total	Year	Quarter	Total
1826	2	1,522	1831	4	1,096
	3	1,562	1832	1	853
	4	1,403		2	865
1827	1	1,297		3	869
	2	1,563		4	877
	3	1,559	1833	1	706
	4	1,478		2	854
1828	1	1,190		3	863
	2	1,462		4	877
	3	1,427	1834	1	778
	4	1,468		2	860
1829	1	1,208		3	891
	2	1,557		4	892
	3	1,561	1835	1	816
	4	1,526		2	1,045
1830	1	1,293		3	1,040
	2	1,499		4	1,062
	3	1,342	1836	1	975
	4	1,309		2	1,108
1831	1	1,120		3	1,111
	2	1,282		4	1,125
	3	1,132			

Year	Quarter	Pickmen	Labourers	Washers	Total
1837	1	726	119	165	1,010
	2	692	110	329	1,131
	3	666	106	335	1,107
	4	681	106	335	1,122
1838	1	739	124	64	927
	2	704	122	311	1,137
	3	702	115	308	1,125
	4	700	114	320	1,134
1839	1	724	135	70	929
	2	699	109	320	1,128
	3	670	111	319	1,100
	4	686	109	299	1,094
1840	1	722	130	75	927
	2	696	111	320	1,127

Year	Quarter	Pickmen	Labourers	Washers	Total
1840	3	704	119	325	1,148
	4	690	111	350	1,151
1841	1	712	130	158	1,000
	2	722	119	357	1,198
	3	738	108	389	1,235
	4	742	99	360	1,201
1842	1	773	117	160	1,050
	2	736	103	359	1,198
	3	740	101	372	1,213
	4	732	105	396	1,233
1843	1	772	121	137	1,030
	2	736	112	358	1,206
	3	702	114	320	1,136
	4	734	107	334	1,175
1844	1	728	112	69	909
	2	719	102	327	1,148
	3	727	106	327	1,160
	4	736	104	326	1,166

The earlier figures are from M. Hughes, *Land, lead and coal*, 1963, based on returns in Greenwich Hospital letters, Adm. 66/105–111. The later figures are from Adm. 79/56.

Glossary

Bing. The standard measurement of lead ore, equivalent to 8 cwt.

Bingtale. A contract for payment according to the weight of the washed ore delivered.

Bouse. Unwashed ore.

Bucker. Hammer for crushing bouse.

Buddle. Apparatus for washing ore.

Dead work. Mining operations intended to open up a vein, not to raise ore.

Deads. Waste material extracted from the mine.

Fathomtale. A contract for payment according to the distance driven.

Flat. Horizontal ore deposit.

Fore-head. The face of a mine working.

Galloway. Either a single pack pony or a team.

Galloway level. A railed level from which horses could drag bouse and deads.

Grove. A lead mine.

Hush. Either the removal of surface soil by a rush of water in an attempt to disclose a vein or the result of such action.

Kibble. A small bucket.

Length. Place of work specified by contract.

Level. A mine adit.

Pickings. Old workings.

Sump. A shaft which does not reach the surface.

Tack. The leasing of a small amount of ground for mining.

Tontale. A contract for payment according to the weight of lead smelted.

Vogue. A waggon used underground.

Washing. The dressing of lead ore.

Wastes. That part of the bouse discarded during washing.

Bibliography

This bibliography lists the principal sources—MS and printed—used in the writing of this study. It includes only material directly about the lead mining region of the northern Pennines. Books about other regions and general economic histories are not included even if cited in the text. Local histories, gazetteers and directories, etc., are listed only if they contain relevant and novel information.

1 *MS sources*

BRITISH MUSEUM DEPARTMENT OF MANUSCRIPTS

Blackett lead mines in Allendale and Weardale. Chief agent's letter book, September 1728 to June 1734. Add. MS 27420. (a 'stray' from the Blackett/Beaumont records now in Newcastle University Library).

PUBLIC RECORD OFFICE

Greenwich Hospital records relating to the administration of the Hospital's northern estates, Adm. 66.
Census papers: enumerator's schedules, 1851.

NEWCASTLE UPON TYNE UNIVERSITY LIBRARY

Blackett/Beaumont lead mining records in Allendale and Weardale, 1723–1888: B/B 1–175.
Hexham manorial papers.
Thomas Bell collection of estate and enclosure records.
Diary of Thomas Sopwith, 1823–79 (a microfilm of the 167 volumes in the possession of Mr Anthony Sopwith, Bradfield College, Berks.).

NEWCASTLE UPON TYNE PUBLIC LIBRARY

Diary of Matthias Dunn, 1831–35.
John Bell collections relative to the estates of the Greenwich Hospital.

NEWCASTLE UPON TYNE LITERARY AND PHILOSOPHICAL SOCIETY LIBRARY

An account of the method of smelting lead ore by James Mulcaster, *c.* 1805.

NORTH OF ENGLAND INSTITUTE OF MINING AND MECHANICAL ENGINEERS LIBRARY, NEWCASTLE UPON TYNE

London Lead Company records, 1692–1899: L.L.C. 1–37.

NORTHUMBERLAND COUNTY RECORD OFFICE, GOSFORTH, NEWCASTLE UPON TYNE

Blackett (Matfen) MSS (include mining records relating to Fallowfield).

Blackett (Wylam) MSS (include papers of Christopher Blackett, chief agent of the Beaumont mines at the end of the eighteenth century).

WIGAN PUBLIC LIBRARY

An account of the method of smelting lead ore by James Mulcaster, *c.* 1805 (another copy of the MS in the Newcastle upon Tyne Literary and Philosophical Society Library, with a slightly variant text).

Greenwich Hospital. Reports on the mines on Alston Moor, 1821, with copies of those made in 1778.

ALLENHEADS ESTATE OFFICE

Various MSS relating to the administration of the Blackett/Beaumont mines and estates.

SETTLINGSTONES MINE

Various MSS and plans relating to mining in Allendale.

HUNSTANWORTH MINE

Derwent Mines Co. Ltd records.

2 *Parliamentary papers*

Journals of the House of Commons.

Decennial census of Great Britain, 1801–1901.

Charities in England and Wales for the education of the Poor. Commissioners. Vol. 21, 1829; vol. 23, 1830. 1829 (349) VIII; 1830 (462) XII.

Administration and practical operation of the poor laws. Royal Commission report, 1834 (especially Appendix B1, *Answers to rural queries* and Appendix A, *Report of Assistant Commissioners*, part 1, 'Cumberland' by Captain Pringle; 'Northumberland and Durham' by John Wilson). 1834 (44) XXVIII; 1834 (44) XXIV.

Minutes of the Committee of Council on Education, 1840–41 (pp. 148–51, 'Statistics having reference to the . . . inhabitants of Alston Moor.')

Children's employment (mines). Royal Commission. Appendix, part 1, report by J. R. Leifchild (abbreviated as '1842 report (Leifchild)'). Appendix, part, 2, report by Dr J. Mitchell (abbreviated as '1842 report (Mitchell)'). 1842 (381) XVI; 1842 (382) XVII.

State of the population, education and schools in the mining districts. H. S. Tremenhere, commissioner. Annual reports, 1844–59 (the commissioner visited Allenheads in 1851).

Geological survey. Mineral statistics (collected) by Robert Hunt, 1848–81.

Rating of Mines. Select committee, 1857 (includes evidence from the chief agents of the Beaumont concern and the London Lead Company). 1857 sess. 2 (241) XI.

Sanitary state of the people of England. Papers, 1858 (paper by E. H. Green-how includes evidence relating to Alston). 1857–58 (2415) XXIII.

Medical Officer of the Privy Council: third report, 1860.

State of popular education in England. Royal Commission report, vol. 2, 1861 (includes A. F. Foster on 'The mining districts of Durham and Cumber-land'). 1861 (2794–II) XXI, part II.

Conditions in mines to which the provisions of the Act 23 and 24 Vict. c. 151 do not apply, with reference to the health and safety of persons employed (2 vols.), 1864 (abbreviated as '1864 report (vol. 1); '1864 report (vol. 2)'; '1864 report (appendix)'). 1864 (3389) XXIV, part 1; 1864 (3389) XXIV, part 2.

Depressed conditions of the agricultural interests. Royal Commission. Digest and appendix to part 1, 1881 (report by John Coleman on Durham describes contemporary conditions in Teesdale and Weardale). 1881 c. 2778–II XVI.

3 *Books and articles in periodicals*

J. W. Allan, *North-country sketches.* Newcastle, 1881.

C. Armstrong, *Pilgrimage from Nenthead.* London, 1938.

A. Aspinall, *The early English trade unions: documents.* London, 1949.

J. Bailey, *General view of the agriculture of the county of Durham.* London, 1813.

J. Bailey, G. Culley and A. Pringle, *General view of the agriculture of the counties of Northumberland, Cumberland and Westmorland.* London, 1813.

J. G. Baker and G. R. Tate, *New flora of Northumberland and Durham.* London, 1868.

F. Brook, 'Fallowfield lead and witherite mines'. *Industrial archaeology*, vol. 4, 1967, pp. 311–22.

T. F. Bulmer, *History, topography, and directory of Northumberland* (*Hexham division*). Manchester, 1886.

W. Cobbett, *Tour in Scotland and in the four northern counties of England in 1832.* London, 1833.

L. C. Coombes, 'Lead mining in East and West Allendale'. *Archaeologia Aeliana*, fourth series, vol. 36, 1958, pp. 245–70.

T. Crawhall, 'An account of certain instruments formerly used for the purpose of blasting in the lead mines at Allenheads'. *Archaeologia Aeliana*, vol. 1, 1822, pp. 182–6.

G. Dickinson, *Allendale and Whitfield*, second edition, Newcastle, 1903.

J. Dickinson, 'Lead mining districts of the north of England'. *Transactions of the Manchester Geological Society*, vol. 27, 1900–02, pp. 218–68.

J. L. Dobson, 'Charitable education in the Weardale district of County Durham, 1700–1830'. *Durham Research Review*, vol. 2, 1955–59, pp. 40–53.

P. A. Dufrenoy and J. B. A. L. L. Elié de Beaumont, *Voyage métallurgique en Angleterre*, second edition, 2 vols, Paris, 1837–39.

K. C. Dunham, *Geology of the northern Pennine orefield*, vol. 1, 1948 (*Memoirs of the Geological Survey of Great Britain*).

—'The production of galena and associated minerals in the northern Pennines' *Transactions of the Institution of Mining and Metallurgy*, vol. 53, 1944, pp. 181–252.

Sir F. Eden, *The state of the poor*, 3 vols. London, 1797.

W. M. Egglestone, *The projected Weardale railway*. Stanhope, 1887.

—*The Weardale nick-stick*. Stanhope, 1872–74.

J. R. Featherston, *Weardale men and manners*. Durham, 1840.

W. Fordyce, *The history and antiquities of the county palatine of Durham*, 2 vols. Newcastle, 1857.

W. Forster, *A treatise on a section of the strata from Newcastle upon Tyne to Cross Fell*, second edition. Alston, 1821. Third edition, 'revised and corrected to the present time by the Rev. W. Nall', Newcastle, 1883.

J. J. Graham, *Weardale past and present*, Newcastle, 1939.

J. Granger, *General view of the agriculture of the county of Durham*. London, 1794.

Greenwich Hospital: [printed] reports and surveys of the estates. London, 1805, 1813, 1815, 1821, 1822.

—Report on the state and condition of the roads and mines on the estates of the Greenwich Hospital, ed. E. H. Locker, 1823.

J. Housman, *A topographical description of Cumberland, Westmorland and Lancashire*. Carlisle, 1800.

R. Hunt, *British mining*. London, 1884.

W. Hutchinson, *The history of the county of Cumberland*, 2 vols. Carlisle, 1794.

G. Jars, *Voyages métallurgiques*, 3 vols. Lyons and Paris, 1774–81.

F. Jollie, *Cumberland guide and directory*. Carlisle, 1811.

J. Lee, *Weardale memories and traditions*. Consett, 1950.

J. Leithart, *Practical observations on mineral veins*. Newcastle, 1838.

Lord Fitzhugh . . . or the parish magazine, localised for both sides of the Tees. Barnard Castle, c. 1880.

J. Losh, *Diaries and correspondences*, vol. 2, *Diary*, 1824–33, ed. Edward Hughes 1959 (Surtees Society publications, vol. 174).

H. Louis, 'Lead mines in Weardale'. *Mining Magazine*, vol. 16, 1917, pp. 15–25.

E. Mackenzie, *A historical and descriptive view of the county of Northumberland*, 2 vols. Newcastle, 1811. Second edition, 2 vols, Newcastle, 1825.

E. Mackenzie and M. Ross, *An historical, topographical and descriptive view of the county palatine of Durham*, 2 vols. Newcastle, 1834.

D. Macrae, 'The improvement of waste lands.' *Journal of the Royal Agricultural Society*, second series, vol. 4, 1868, pp. 321–34.

F. J. Monkhouse, 'The Greenwich Hospital smelt mill at Langley, Northumberland, 1768–80'. *Transactions of the Institution of Mining and Metallurgy* vol. 49, 1940, pp. 701–9).

—'The transport of lead and silver from Langley Castle to Newcastle, 1768–1779'. *Proceedings of the Society of Antiquaries of Newcastle upon Tyne*, fourth series, vol. 9, 1939–41, pp. 176–86.

W. Nall, 'The Alston mines'. *Transactions of the Institution of Mining Engineers*, vol. 24, 1902–03, pp. 392–412.

G. Neasham, *North-country sketches*. Durham, 1893.

J. G. Nevin, 'The general system of farming in Alston, Allendale, Weardale, and the south-west part of Hexhamshire'. *Proceedings of the Durham College of Science Agricultural Students' Association*, vol. 1, 1900–01, pp. 21–31.

Newcastle and Carlisle Railway Bill. Copy of the evidence taken before a committee of the House of Commons. Newcastle, 1829.

Newcastle upon Tyne Sunday School Union, fifth report. Newcastle, 1823.

J. Nicholson and R. Burn, *The history and antiquities . . . of Westmorland and Cumberland*, 2 vols. London, 1777.

W. Parson and W. White, *History, directory and gazetteer of the counties of Durham and Northumberland*, 2 vols. Leeds, 1827.

—*History, directory and gazetteer of Cumberland and Westmorland*. Leeds, 1829.

W. M. Patterson, *Northern Primitive Methodism*. London, 1909.

H. L. Pattinson, 'An account of the method of smelting lead ore and refining lead, practised in the mining districts of Northumberland, Cumberland, and Durham, in the year 1831'. *Transactions of the Natural History Society of Northumberland, Durham and Newcastle upon Tyne*, vol. 2, 1831, pp. 152–77.

J. Percy, *The metallurgy of lead*. London, 1870.

R. Pococke, *Northern journeys of Bishop Richard Pocoke (1760)*, 1914 (Surtees Society publication, vol. 204).

Practical illustrations of the benefits to be derived from well-conducted friendly societies, with reference more particularly to the new societies established in East and West Allendale. Newcastle, 1852.

A. Raistrick, 'Lead smelting in the north Pennines during the seventeenth and eighteenth centuries'. *University of Durham Philosophical Society Transactions*, vol. 9, 1931–37, pp. 164–79.

—'The London Lead Company, 1692–1905'. *Transactions of the Newcomen Society*, vol. 14, 1933–34, pp. 119–62.

—'Ore dressing in the eighteenth and early nineteenth centuries'. (*Mine and Quarry Engineering*, 1939, pp. 161–6).

—*Two centuries of industrial welfare: the London (Quaker) Lead Company, 1692–1905*. London, 1938.

A. Raistrick and B. Jennings, *History of lead mining in the Pennines*. London, 1965.

B. W. Richardson, *Thomas Sopwith, with excerpts from his diary of fifty-seven years*. London, 1891.

G. Ritchell, *An account of certain charities in Tyndale ward in Northumberland*. Newcastle, 1713.

W. Robinson, *Lead miners and their diseases*. Newcastle, 1893.

A. E. Smailes, 'The lead dales of the northern Pennines'. *Geography*, vol. 21, 1936, pp. 120–9.

T. Sopwith, *An account of the mining districts of Alston Moor, Weardale, and Teesdale*. Alnwick, 1833.

—*Observations addressed to the miners and other workmen employed in Mr Beaumont's lead mines*. London, 1846.

—*Substance of an address to the members of the St John's Chapel Friendly Society*. Hexham, 1847.

T. Sopwith junior, 'The dressing of lead ores'. *Minutes and Proceedings of the Institution of Civil Engineers*, vol. 30, 1869–70, part 2, pp. 106–35.

Spar from the high flat: a Christmas annual for Allendale and surrounding dales, No. 1. Allenheads, 1871.

A. Steele, *History of Methodism in Barnard Castle and the principal places in the dales circuit*. London, 1857.

Teesdale Mercury, Tales and traditions, 3 parts. Barnard Castle, *c.* 1885.

W. Turner, 'Account of a short tour through the lead mine districts, in 1793'. *Transactions of the Literary and Philosophical Society of Newcastle upon Tyne*, vol. 1, part 1, 1831, pp. 66–81.

'Visit to the Stonecroft and Greyside lead mines . . . and the Settlingstones lead mines, 19 October 1877'. *Transactions of the North of England Institute of Mining and Mechanical Engineers*, vol. 27, 1877–78, pp. 15–22.

W. Wallace, *Alston Moor: its pastoral people, its mines and miners*. Newcastle, 1890.

—*The laws which regulate the deposition of lead ore in veins*. London, 1861.

J. Wallis, *The natural history and antiquities of Northumberland*, 2 vols. London, 1769.

J. Walton, 'The mediaeval mines of Alston'. *Transactions of the Cumberland and Westmorland Antiquarian and Archaeological Society*, new series, vol. 45, 1945.

R. Watson, *Poems and songs of Teesdale*. Darlington, 1930.

E. Welbourne, *The miners' unions of Northumberland and Durham*. Cambridge, 1923.

'W. F. S.', Methodism in Allendale. *Methodist Magazine*, 1872, pp. 710–19.

W. Whellan, *History, topography and directory of Durham*. London, 1856.

W. White, *Northumberland and the border*. London, 1859.

P. N. Wilson, 'The Nent Force level'. *Transactions of the Cumberland and Westmorland Antiquarian and Archaeological Society*, new series, vol. 63, 1963, pp. 253–80.

4 *Theses*

M. Hughes, *Lead, land and coal as sources of landlord income in Northumberland between 1700 and 1850*. Durham University (King's College, Newcastle) PhD thesis, 1963.

A. E. Smailes, *The dales of north-east England*. London University MA thesis, 1932.

D. Thompson, *The rural geography of the west Durham Pennines: Derwentdale, Weardale and upper Teesdale*. Manchester University MA thesis, 1962.

Index

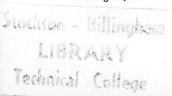